The Unreliable Nation

Inside Technology

edited by Wiebe E. Bijker, W. Bernard Carlson, and Trevor Pinch

For a list of titles in the series, see the back of the book.

The Unreliable Nation

Hostile Nature and Technological Failure in the Cold War

Edward Jones-Imhotep

The MIT Press
Cambridge, Massachusetts
London, England

Set in StoneSerif by Westchester Book Composition. Printed and bound in the United States of America.

Library of Congress Cataloging-in-Publication Data

Names: Jones-Imhotep, Edward, author.
Title: The unreliable nation : hostile nature and technological failure in the
  Cold War / Edward Jones-Imhotep.
Description: Cambridge, MA : MIT Press, [2017] | Series: Inside technology |
  Includes bibliographical references and index.
Identifiers: LCCN 2016054825 | ISBN 9780262036511 (hardcover : alk. paper)
Subjects: LCSH: Technology and state—Canada—History—20th century. |
  Telecommunication policy—Canada—History—20th century. | National
  security—Canada—History—20th century. | Telecommunication systems—
  Canada, Northern—Reliability—History—20th century. | Cold War. |
  System failures (Engineering)
Classification: LCC T23.A1 J66 2017 | DDC 338.971/0609045—dc23
LC record available at https://lccn.loc.gov/2016054825

10   9   8   7   6   5   4   3   2   1

for my parents

Contents

and steel.[5] The examples can be repeated again and again. As analogies for the nation, as foils in its historical development, as ways of thinking about historical necessity and freedom, nationalism after nationalism has enlisted both natures and technologies to solidify their credentials and to appeal to an authority beyond the seemingly transitory qualities of the social, the cultural, and the political.[6]

Only recently, however, have historians begun investigating the practices that helped portray nations themselves as natural and technological objects in the first place. They have generally taken one of two separate paths. Historians of science have examined how scientists and natural philosophers catalogued the rich and varied natural worlds that helped cast individual nations as unique collections of flora, fauna, and landscapes, making them appear transcendent, legitimate, and even inevitable.[7] Historians of technology have instead focused on the role of nation-building technologies, particularly the large, highly visible infrastructure projects that politicians and technologists transformed into potent symbols through complex negotiations and political theater.[8] But what we are missing, and what episodes like the one at the center of this book provide, is an understanding of how those two powerful ways of apprehending the modern nation—the natural and the technological—mapped onto and shaped each other around *failing* machines: how the practices and ambitions that crafted the nation as a natural object were allied with attempts to understand it as a technological space, and specifically as a territorial grouping of specific types of machines and their worrying behaviors.

Like many other postwar technoscientific projects, the efforts to secure Northern communications in Canada were part of an attempt to reimagine the nation and its place in the postwar world.[9] The Cold War transformed the circumpolar regions (along with the deep oceans and outer space) into "hostile environments" charged with political, military, and physical threat. They became arenas for a global struggle over territory, knowledge, and strategic advantage.[10] That new, polar-oriented world placed Canada in a particularly vulnerable position: wedged geographically between the superpowers, and pulled politically and culturally between a declining Europe and a rising United States.[11] Within that context, government officials and defense scientists seized on the radio failures of the North as a way of leveraging technological and geographical vulnerability into political power, influence, and distinctive identity. Within that project of a nation trying to reimagine itself, the laboratories of Canada's Defence Research Telecommunications Establishment (DRTE) set out to articulate the nation's status as both a natural object defined by distinctly "Northern" upper-atmospheric phenomena and a technological space of uniquely powerful, widespread radio failures that threatened the technological integrity of the nation. In their most severe form, those phenomena simulated the effects of nuclear detonations in the upper atmosphere, blacking out long-distance radio circuits, disrupting radars, and masking incoming ballistic missiles, making them appear to be the natural or the manufactured precursors of Soviet attack and nuclear war. But even in weaker forms, those phenomena routinely cut off the Canadian North from reliable communications with the rest of the country. For Cold War strategists, their effects seemed to invite Soviet incursions and American occupation and to threaten Canada's survival as an independent entity. Over the first two decades of the Cold War, the scientists and engineers of DRTE wove this hostile nature and its failing machines into a sophisticated, anxious, long-standing debate over the relationship between nature, technology, and nation in Canada.

Like all natures, the one these scientists invoked had a complex history.[12] In Western cultures, the Northern regions have historically been a place of both inspiration and dread: the lands of darkness, desolation, and death; but also of hope, opportunity, vitality, and wonder. Those complex associations can be traced partly to the mythical *ultima Thule* of ancient Greek geographies—the most distant place on Earth, the end of the knowable world. But they have deep roots in Enlightenment notions of the sublime and its inseparable union of the noble, the wondrous, and the terrifying.[13] Historically, Northern nature represented a particularly potent and pure embodiment of nature's authority, either indifferent or hostile to human will.[14] Perhaps because of this, scientific epistemologies had long privileged the North as a site of knowledge, a place where nature spoke more directly and more forcefully than in temperate regions.[15] Carried forward through arctic expeditions and magnetic crusades, these epistemologies were vital to the natural philosophy of the eighteenth century and the professionalizing science of the nineteenth and the twentieth centuries, where they underwrote gendered and racialized theories of geographical determinism while furnishing a way for otherwise marginal scientists to make themselves culturally central.[16] Since early in its colonial history, Canada participated in those associations—first to distinguish itself from Britain and France, then in defining its own internal relations, replacing the West with the North as a frontier that framed the fate of the nation.[17] By the middle of the twentieth century, as the Cold War enveloped the polar regions, the idea of the North in Canada was an accretion of these historical "norths," at once natural environment, scientific object, Native homeland, physical landscape, cardinal direction, geopolitical region, and metaphysical force.[18] And in that polyvalence, the North captured an ambiguity about the very concept of nature and its relationship to technologies; one that goes to the heart of how we think and write about their histories.

One common understanding of nature sees it as synonymous with *natural environment*—the nonhuman world of climate, geology, plants, animals, microbes, and other organisms, and their interrelations in natural systems. The elements of this "nature" can change over time, often as a result of human action, making it a primary focus of environmental history and of much of the history of technology.[19] The late-twentieth-century observers Northrop Frye and Margaret Atwood, writing at the endpoint of the story told here, identified this nature as one of the most powerful themes in Canadian culture and history.[20] "Everything that is central in Canadian writing," Frye would claim, "seems to be marked by the imminence of the natural world." That imminence—the brute realities of long harsh winters, vast boreal forests, antediluvian granite, tundra, and taiga—formed what Frye called the "obliterated environment" in Canadian history. Even as it defined the nation, this hostile nature inspired a persistent cultural preoccupation with both human and national survival—a theme that would gain special purchase during the Cold War.[21] Against this nature, technologies figured as antagonists: instruments for controlling the natural environment, exploiting it, amending it, disrupting it, and mastering it physically and cognitively. Aligning themselves with the conservation movements of the 1960s, histories of technology and early histories of the environment would mirror this concern with nature transformed and opposed by technologies; more recent "envirotech" approaches have challenged that opposition while continuing to treat nature and environment as near synonyms.[22] For Frye, though, the opposition was particularly central to Canadian history: "The enormous difficulties and the central importance of communication and transport, the tremendous energy that developed the fur trade routes, the empire of the St. Lawrence, the transcontinental railways, and the north-west police patrols have given [technology] the dominating role in the Canadian imagination."[23] Alongside the idea of a hostile Northern nature, then, stood a mythology surrounding the promise of technology, and particularly communications, for national survival.[24] Although similar visions would be repeated in nations around the world, Frye and other midcentury Canadians would claim that technology made Canada *uniquely* possible, holding it together against the centrifugal forces of an environment and a geography that constantly threatened to tear it apart. Put differently, technologies were what allowed a hostile Northern environment to define the nation without destroying it.

Within those reflections, though, there was another, older understanding of nature at work: nature as *natural order*—the inherent forces of the natural world and the entities and relations they generate and control.[25] Ideas about the character of that order have changed over time, from a highly organized but exceptionable natural order of the European Middle Ages and the Renaissance, for example, to a uniform and inviolable natural order of the moderns.[26] But undergirding them, and others, was the idea of an underlying power producing and regulating the phenomena of the universe and the properties and characteristics of its objects. Within science in particular, natural orders represent a kind of working ontology. If pressed on the range of phenomena that exist in the world, scientists would admit to all kinds of entities. Taken in all their richness, though, natural objects are too plentiful and idiosyncratic to form the basis of a science.[27] For the purposes of their work, scientists construct conceptual worlds populated by groups of uniform working objects and relations that help to explain natural phenomena and to intervene in them. Individuals, random variation, competition, and species in Darwinian evolution; forces, masses, distances, and time in Newtonian physics. Like the natural environments they help explain, these natural orders have local expressions—distinctive wind patterns off the Western Cape, for instance, or peculiar geological formations near the Gulf of Tonkin. Although their entities and relations change in quality or composition over time—they have histories—they are, at any given moment, taken to be transhistorical. Unlike environments, they are untouched by human action. But they can be harnessed in the form of technologies. Because of this, natural orders have historically enjoyed a sympathetic relationship with machines. Since they claim to identify deep regulating forces and natural laws, their entities and relations form the stripped-down, practical microworlds that are manipulated, exploited, and "applied" in working technologies.[28]

This is the "nature" that has most concerned historians of science. One way of retelling the history of modern science and technology is as a deployment of this second "nature" against the first—the use of increasingly abstract and universal natural *orders* to conceive of and fashion technologies that act against natural *environments*.[29] We should be careful about drawing that distinction too sharply or too generally: the two understandings have common roots, and in certain periods and specific fields, they blend into each other kaleidoscopically.[30] Indeed, one of the aims of this book is to show how technological developments can transform natural orders into natural environments. But their different historical relationships to technology—antagonism versus sympathy—and the distinct ways they are invoked to support the nation make it useful to keep them analytically separate. Whereas nineteenth-century Canadians, for example, often saw specific environmental features—muskeg, tundra, frigid temperatures—opposing the creation of the nation, they could see the overarching natural orders of geology, meteorology, thermodynamics, terrestrial magnetism, and botany supporting the idea of a transcontinental nation and making its historical emergence inevitable.[31] Like local flora, fauna, and landscapes, natural orders also contribute to place-making and identity. But whereas landscapes and environments can be altered and even lost, natural orders are often taken as inexorable. They derive their power from the way they abstract and transcend, describe and authorize—the way they mix "is" and "ought."[32] Historically, they have naturalized nations in ways that were presented as more transcendent, inevitable, and essential than the landscapes and environments that form their epiphenomena.

One of the central arguments of this book is that the project of extending reliable communications to the Canadian North formed one half of a twentieth-century realignment of technology, environment, and natural order in Canada. In the first half of the century, influential commentators including the historian Harold Innis had taken the traditional antagonism between environment and technology and turned it into a deep sympathy that explained both Canada's history and its future.[33] Looking back from the 1930s, Innis envisioned Canada emerging from the successive exploitation of a series of geographically determined "staples" made possible by the St. Lawrence River (along which the staples moved) and by a set of technologies attuned to those natural systems (for example, the canoe). For Innis, that entrainment between specific environmental features and particular technologies gave the nation its character and identity—a society of semi-nomadic traders and trappers rather than farmers, alliances rather than wars with Native peoples, westward expansion in search of animal pelts, and so on.[34] Asserting his nuanced brand of geographical determinism, intimately tied to technology, Innis asserted that Canada was created "not in spite of geography but because of it."[35] The force of the DRTE communications project lay in a parallel move. Innis had redrawn the age-old antagonism between technology and *environment* by aligning them in support of the nation; the work of postwar defense scientists took the age-old sympathy between technologies and *natural orders*, recast them as antagonists, and located their point of opposition in the problematic behavior of machines that threatened Canada's survival.

They did this by elaborating a series of "machinic orders"—functional groupings of machines structured around concerns about their collective

behavior and its variation across time and place. When DRTE started its
work, neither the spatial extent of communications disruptions nor their
causes were immediately obvious. To identify them, the laboratories needed
to assemble individual instances of radio failures and malfunctions into a
broader understanding of where those behaviors were located in relation
to natural phenomena and how they varied under changing conditions.
These mid-level understandings of failure—focused neither on individual
machines, nor on every instance of the same machine—were part of a more
general historical development. Just as one of the ambitions of modern
states has been to map and order the characteristics and spatial distributions of people, throughout the nineteenth and twentieth centuries nations
also struggled to map and understand their territory as a technological
space: part of the larger attempts to create a coordinated image of the nation
that mixed the mapping of people and the mapping of "things."[36] In those
understandings, the number and location of certain types of machines—
steam engines, early computer technologies, short-wave radios, nuclear
reactors—mattered. And with the rise of techno-nationalism, aspects of
their behavior especially came to matter in ways they had not before.[37]
As nations developed, then, they embraced a preoccupation for treating
both technologies and technological effects over their entire terrain, making
them legible in ways that supported the classic political responsibilities
of the state, including the regular circulation of people, goods, and information.[38] These were not "systems" in the sense of heterogeneous elements
assembled and directed toward coordinated and concrete ends.[39] They were
instead malleable typologies of "similar" machines that embodied concerns about their functioning and inspired attempts to map, monitor, and
intervene in their behavior. It is useful to think of these functional and
spatial understandings as "machinic," rather than "technological," because
"machinic" connotes a specificity or concreteness that is useful—specific
machines, concrete behaviors—while avoiding the connotations of claus-
trophobic determinism and autonomous historical force that the term
"technology" sometimes carries.[40] I call their coordinated representation
"orders" because of the parallels with the natural. If natural orders draw
together diverse but interrelated entities and phenomena behind ordered
explanations of the natural world, then the machinic orders of the modern
period provided ordered understandings of the way specific types of machines
occupied and acted in that world.

A main focus of this book is to show *how* DRTE linked the natural and
machinic orders, using their relationship to represent the vulnerability of

hindsight, are defined by their unrealized potential—a meaning that closely tracks the nineteenth-century origins of "failure" as a kind of person (originally a type of man).[48] Instead of individual material objects, they represent archetypes—the perpetual-motion machines, atmospheric railways, and picturephones that round out the rich diversity of technological history.[49]

Although technological failures in this sense populate this book—the most important being the project of pan-Northern shortwave communications itself—my main interest is in a second, more pragmatic understanding of failure: not a *class* of technologies, but a *condition* that machines experience. Not failed machines, but failing machines. The failures that interest me are entropies—tendencies toward disorder, degradation, breakdown, surprise, even calamity. Failures in this other sense represent the all-too-often-realized potential in all technologies: events and episodes that arrive out of the blue, with little warning and at inconvenient times, and that are embedded in their own deep webs of social relations.[50] They afflict individual devices and specific systems rather than second-order typologies; they are concrete and material rather than analytical and abstract; and they are generally recognized in the moment by historical actors rather than attributed by posterity. Just as importantly for our purposes, these are the failures that are investigated, assembled, mapped, and correlated with other, worldly phenomena to form machinic orders. They represent a more fine-grained understanding of failure that is particularly useful in understanding the relations of nature and technology in the Canadian North during the Cold War. By almost any measure, shortwave radio was no "failed" technology; but in specific places, at specific times, in the presence of certain natural phenomena, radio sets and related technologies failed to work as intended.[51] These localized failing machines might, as a historical contingency, build up to produce failed technologies. But that was far from inevitable. By treating all technological failures—from failed mega-projects to minor technical malfunctions—under the same name, we blur the different and even orthogonal historical questions they raise.

This second class of failures has much to teach us about the role of technology in defining the characteristics, the capabilities, and even the agency of nature, as the history of natural disasters suggests.[52] Natural disasters have historically been episodes of conceptual and epistemic, as well as physical, damage. The earthquake that struck Lisbon in the late fall of 1755 shook not only buildings but also spiritual faith, inspiring Voltaire's famous rejection of theodicy and unsettling the place of evil in the world.[53] More generally, the violence, unpredictability, and unruliness of earthquakes, their ability to escape established categories of thought and practice, have repeatedly opened up fissures in structures of knowledge and communities of knowers.[54] But alongside knowledge and beliefs, the behavior of technologies has shaped the very definition and possibility of natural disasters. Far from being the mere casualties of those events, material cultures help create and define different registers of the catastrophic, specifically the natural and the social. The disaster of Hurricane Katrina was generated not just by a powerful tropical cyclone in the Gulf of Mexico but also by a neglected infrastructure and an ecology of poverty, under-privilege, and inequality in New Orleans.[55] Natural catastrophes are neither unproblematically "natural" nor unambiguously "catastrophic"; their catastrophic character is made possible by social, political, and economic determinants of vulnerability.[56] We can add the technological to that list. Buildings, bridges, and levees are technologies, and it is precisely the discriminate damage to them—their large-scale but varying destruction, the way they shield and protect but also shatter and extinguish lives when they crumble and collapse—that translates the natural event into social, political, and economic cataclysm. Historically, then, technologies are not simply the instruments through which humans act on natures or inscribe them; they are not merely tools or devices. They are also the media through which the natural gains force, specificity, and even agency in reshaping societies and polities, sometimes with devastating effects.[57] Out of those effects and in the wake of those disasters, historical actors craft identities, advantages, and meanings. Out of the geological, social, and conceptual fissures opened up by earthquakes, eruptions, tsunamis, and typhoons emerge opportunities for people to reimagine their world, including the relationship between natures and machines.

The historical actors discussed in this book dealt with nature at a scale and with a power and an unpredictability that rivaled the catastrophic. The disturbances they tracked circled the globe—auroral displays and magnetic storms so powerful and widespread they could reverse local magnetic fields at the earth's surface, darken power grids, and blind radar defenses, mimicking the effects of natural disasters. Overwhelmingly, though, these scientists brought together nature, machines, vulnerability, and political advantage around more humble phenomena that, when articulated through scientific images and combined with political maps and Cold War strategems, became symbolically powerful. Those failures were used to define one of the preeminent hostile natures of the Cold War—the Canadian North—and they had analogues throughout the history of technology. Decay, degradation, wear, cracking, and corrosion are also "natural" processes. Rather than opposing nature or blending seamlessly with it, technologies have figured orders and machine behaviors helped to clarify the threats that drove the Cold War and to create the understandings of hostile natures and technological failure that sustained it.[69]

In this way, the book addresses broader questions about the period. How did nations mobilize not just military forces and strategic alliances, but also natural phenomena and machine behaviors behind survival, autonomy, and identity? How did the exploration and mapping of natural and machinic orders shape the geopolitical order of the Cold War and the perceptions of threat on which it depended? And how were the period's pressing anxieties about failing machines part of a longer natural history of hostility in the twentieth century? If the Cold War was at first a contingent and later a more systematic attempt to project the United States or the Soviet Union into the world, then a central preoccupation for many nations was how to engage those projections effectively, carving out spaces of reluctant collaboration, outright resistance, and hard-fought autonomy against their homogenizing tendencies. Nature and machines formed some of the most powerful resources for prosecuting the Cold War; but they also provided some of the most authoritative means of escaping its totalizing tendencies.[70] In these actionable spaces between nature and machines, contemporaries saw possibilities for resistance, distinction, and identity. Exploring those global struggles moves us beyond the conceit that the only stories to be told are those centered, directly or by proxy, on Washington or Moscow, or on the classic technologies of the conflict. In that sense, the story told here shows how a nation caught geographically between two superpowers—a country struggling with "an uncertain identity, a confusing past, and a hazardous future"—could speak beyond itself.[71]

Structure of the Book

The book develops along three main lines. Chapters 1–3 focus on DRTE's investigations of the natural and machinic orders implicated in Northern shortwave radio communications. Beginning with World War II, the chapters explore how upper-atmospheric phenomena were interpreted through the behavior of machines, how Canadian researchers recast wartime radio failures in the North Atlantic as a problem of the high-latitude ionosphere above Canada, and how those researchers worked to articulate a distinctive natural order through the production and interpretation of ionograms.

Chapters 4–6 examine attempts to circumvent that specific relationship between nature and machines while keeping its powerful symbolism and its political importance intact. They trace new relationships between natures and machines through a series of projects—ionospheric satellites that recast the natural order as hostile environment, schemes for mass producing and mass interpreting the images those satellites produced, and communications technologies designed to adapt automatically to the turbulent and characteristic high-latitude ionosphere or to bypass it altogether. In doing so, those chapters illustrate how DRTE's own work eroded the connections between natural and machinic orders it had worked so hard to create.

Chapter 7 steps back to show how DRTE's making and unmaking of that relationship between nature and technology intersected with broader changes in the technological and human geography of the Canadian North over the two decades after World War II. It explores how the complex interaction of nature, technology, and human populations came to define the Canadian North through failed communications. The book closes with some reflections on how those developments can help us rethink our histories of nature and technology in the modern period.

1 The Nature of War

World War II instrumentalized nature.[1] Across a wide range of activities and disciplines, researchers struggling to understand and improve the performance of wartime technologies continually recast natural phenomena in terms of the machines that consumed their thoughts and actions. Working on projects ranging from signal distortion to atomic bombs, the physicist Richard Feynman developed a powerful and pragmatic theoretical culture built around representations of the sub-atomic world as resolutely modular as the macroscopic machines they made possible.[2] The psychologist B. F. Skinner, musing about problems of missile guidance to thwart the devastation of German bombing raids, saw pigeons as promising guidance "instruments," not because they were intelligent, but because they could be "made into a machine, from all practical points of view."[3] The polymath Norbert Wiener, scrambling to provide a defense against the same German aerial onslaught in the late summer of 1940, modeled the actions of both Allied gunners and enemy pilots as self-regulating servomechanisms.[4] For all these historical figures, the functioning and malfunctioning of machines said something both literal and metaphorical about the *natural* world. And they were not alone. In discipline after discipline, nation after nation, problem after problem, researchers translated the natural phenomena at the core of their research into languages and concepts born out of the world of machines.[5]

This is the "nature" of war that interests me here—the visions of natural phenomena built around wartime machine behaviors, specifically the performance of short-wave radio technologies used in global warfare. Those technologies turned high-frequency radio signals into weapons against the enemy.[6] Used for trans-oceanic and intercontinental communications, electronic espionage, and detection, they were built around a working understanding of the ionosphere—the ionized regions of the upper atmosphere that make long-distance short-wave communications possible.[7] Straddling geophysics and radio engineering, shot through with commercial, military and political anxieties, the ionosphere and its phenomena were continually parsed, understood, and represented through the collective behavior of groups of machines. Mapped globally, those behaviors would become the basis of the sweeping machinic orders of wartime communications: abstract and powerful understandings of the spatial distribution of machines and machine-based effects. In order to understand and control those behaviors, wartime researchers cast ionospheric phenomena as a stripped-down "nature" that transformed ion distributions and atmospheric dynamics into weapons against the enemy. That nature, I argue, would form the nucleus of a combined vision of nature and technology central to geopolitics and national identity in Canada after the war.

This chapter tracks three transformations in this "nature" of war, making three interrelated arguments. First, the chapter argues that the anxiety over the reception and detection of short-wave signals during World War II led researchers to technologize the geophysical phenomena that guided signals across oceans and continents. Particularly in what we might call the Sciences of the Enemy—electronic espionage, radio traffic analysis, code-breaking, counterintelligence, and anti-submarine warfare—radio researchers operationalized ionospheric phenomena, defining natural features in terms of their large-scale effects on machines. Although this link between natural order and machine behavior began as a general link between *global* geophysics and radio performance, the chapter's second argument is that the association quickly became highly specific. The unexpected discovery of an atmospheric anomaly above India in early 1943 encouraged Allied researchers to connect the global behavior of machines to a particularly turbulent *polar* ionosphere and geophysics. As the focus of the war shifted to the Aleutian Islands and the Pacific in the summer of 1942, British, American, and Canadian researchers launched a large-scale expansion of the field-stations used to investigate atmospheric phenomena and focused attention on Canadian high-latitude stations to learn more about the polar ionosphere and its effects on radio disruptions. It was in this context, the chapter finally argues, that a Canadian research group charged with signals research and anti-submarine warfare translated this polar relationship into a set of characteristic *national* effects. As part of a postwar project to emphasize Canada's "northernness," they mapped machine behaviors and atmospheric phenomena in an attempt to argue that a distinctive Canadian natural order was the source of global radio disruptions, and to weave that relationship into a broad national mythology surrounding the North and communications. In doing so, they turned the wartime operationalization of the ionosphere on its head: naturalizing technological failures, reading specific natural orders into the erratic behavior of machines. That overall trajectory—from global to polar to national—forged a relationship between nature and technology in which the failures of machines could simultaneously say something about turbulent nature, technological failure, and the postwar nation.

Images and the Enemy

The key to understanding those transformations lies in what we might call the Sciences of the Enemy—the technical activities supporting surveillance, intelligence, and counterintelligence operations during World War II. Unlike the activities immediately involved in engaging and destroying enemy forces, these sciences deployed a set of practices aimed at representing the presence and threat of adversaries, piecing together and understanding their communications, dispositions, and deployments. Cutting across the "Manichean Sciences" of cybernetics, operations research, and game theory, the Sciences of the Enemy furnished the representations these other fields would ultimately analyze and act upon: the location of enemy submarines; the deployment of Axis troops; the number, position, and heading of enemy planes and warships.[8] Physically and symbolically occupying the spaces just outside the "black rooms" of Allied intelligence and the German *Abwehr* (the military intelligence service) their media included cathode-ray displays, radio traffic analyses, submarine plots, and maps. But because their work depended so heavily on short-wave communications of all kinds, they also relied on a set of antecedent representations, the most important of which were known as ionograms.

Ionograms were radio-echo graphs of the electrified regions of the upper atmosphere known as the ionosphere—the shell of electrons and ions that encircles Earth between the stratosphere and near space, stretching from about 50 to about 1,000 kilometers above the planet's surface and changing shape and size with daylight, the seasons, and solar events. The idea that the atmosphere might contain electrified regions began as a hypothesis, proposed by the German Romantic mathematician Carl Friedrich Gauss around 1839 to account for observed variations in Earth's magnetic field. Supported by turn-of-the-century developments that included Marconi's successful trans-Atlantic experiment, scientists began to suggest these electrified regions were composed of discrete ionized "layers."[9] Given the right density, these layers of ionized particles became opaque to radio waves, reflecting them back to the ground. Together, Earth and ionosphere formed an enormous waveguide, channeling radio transmissions over oceans and landmasses to points far beyond the horizon. Spurred by commercial and military interests after World War I and by independent experiments that seemed to "detect" these layers around 1924, atmospheric researchers began attempting to represent the conducting regions visually, in the form of graphs.[10]

The instrument that produced the graphs emerged from the same material and practical culture that produced radar. After experiments conducted in 1925 and 1926 confirmed the existence of conducting ionization in the upper atmosphere, British researchers at Ditton Park began elaborating a powerful and flexible material and practical culture at the intersection of pulse circuitry, oscillography, and graphic techniques including photography. One stream of that research is well known: after rejecting the feasibility of a "radio death ray," Ditton Park's superintendent, Robert Watson-Watt, began modifying the direction finders that were used to measure the height of the ionosphere. He crafted a succession of machines to detect the arrival of atmospherics—radio disturbances produced by naturally generated radio waves, in this case by lightning strikes. Adding oscilloscope displays to the devices allowed him to display the waveform and duration of the disturbances. Eventually adapting those sets and pointing them at planes (instead of thunderstorms), Watson-Watt would go on to produce the radar sets that would be so critical in winning the war.[11]

The radar efforts had an important parallel. Drawing separately on developments in the United States, a second group at Ditton Park mobilized these same materials and sounding techniques to develop machines that helped visualize the ionospheric "layers."[12] In place of enemy planes, these instruments took the ionosphere as their target. By sending varying high-frequency signals skyward and displaying the resulting echoes on an oscilloscope and ultimately on photosensitive paper or film, researchers could translate visual information from the echoes into a plotted graph showing the height of the reflection surfaces as a function of radio frequency—an ionogram.[13] (See figure 1.1.) Later ionosondes would automate this process and produce the graphs directly. Whereas research on radar would quickly focus inward, on the technical details of the apparatus, ionospheric research continued the tradition that invented the ionosonde by focusing on the phenomena represented beyond the experimental spaces of the radio hut or the radar van. The focus of the numerous ionospheric laboratories created during the early years of World War II would be on the flickering display of the oscilloscope and on the resulting images that would be as instrumental in fighting the enemy as the ones produced by radar.

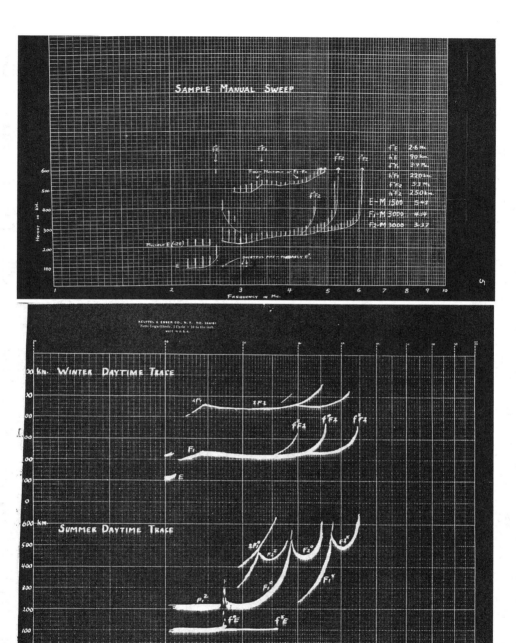

Figure 1.1

Early ionograms. In manual ionosondes, observers measured the time delay of radio echoes (represented as a "pip" on the cathode-ray display or on photographic paper) and plotted their half-heights for given frequencies (upper image). Automatic ionosondes automated this process by directing the cathode-ray traces onto moving film (lower image). Source: Canadian Radio Wave Propagation Committee, "Instructions for Observers: Canadian Ionospheric Stations," July 1944, Library and Archives Canada, Record Group 24, file 9147–2. © Government of Canada. Reproduced with the permission of Library and Archives Canada (2016).

Those images quickly formed the core of a powerful visual language that straddled a natural order of geophysical phenomena on one side and radio communications on the other. A century before the invention of the iono-gram, fostered by developments in aquatints, engravings, and lithographs, geologists had developed methods that allowed them to "read" structural features of Earth's crust often inferred from topography and geological the-ory.[14] The ionogram, growing out of developments in pulse circuitry and photography, turned the long-standing structural interests of geophysics skyward. It seemed to present the ionosphere in cross-section. Like the geo-logical section, which was closely linked to the practical and commercial activities of nineteenth-century mining and quarrying, scientists quickly put the structural insights from the ionogram to work. They devised intri-cate mathematical tools for extracting clues about the natural order that characterized the ionosphere in particular (clouds of electrons and ions, bombarding solar particles, cosmic-ray showers, magnetic forces), and they elaborated methods for linking those phenomena to the actual and pre-dicted performance of radio circuits. By 1934, an international convention had agreed to define the ionosphere as "that part of the upper atmosphere which is ionized sufficiently to affect the propagation of wireless waves," making the performance of radio technologies essential to the definition of the region itself.[15]

Under the urgency of World War II operations, researchers used iono-grams to collapse the complex and lengthy links between natural phenom-ena and machinic behavior at every scale. Where modularity guided the wartime work of the Radiation Laboratory at MIT or the Manhattan Project, wartime ionospheric research was driven by notions of hybridity.[16] It sought to assemble machines and atmospheric phenomena into hybrid systems that fused radio sets and meteorological conditions into stable and pre-dictable communications circuits. Manuals instructing technicians on how to use ionospheric data portrayed the ionospheric layers and Earth's surface as components of a giant waveguide, analogous to the metal ducts used to channel radar pulses in aircraft. They deployed the ionogram to define ionospheric "control points," reflection sites that characterized the upper atmosphere using the language of wartime aircraft manufacturing and ship design. A treatise published in 1945 couched the relation between radio waves and ionosphere in the language of intentionality that would later characterize cybernetics: "Each [radio] ray acts as if it had intelligence and purpose. It bores into the [ionospheric] layer, seeking an electron density sufficiently great to turn it back (by total internal reflection). If such a density does not exist in the E layer at the time, the ray passes through the

F layer, where it repeats its search. If again unsuccessful, the ray passes out into interstellar space."[17] Newbern Smith, head of the US propagation laboratory, contributed to this fusion of the natural and the technological by developing one of the most influential graphic technologies of wartime communications—a transparent overlay whose curved lines identified usable radio frequencies at the point they became tangent to the curves of the ionogram.[18] In a practice repeated thousands of times all over the world, this simple graphic technology preserved the "natural" phenomena on the image while requiring technical staff to read those phenomena through the overlaid operation of machines.

These intensely local practices—laying transparencies over time- and place-specific ionograms—ran parallel to mid-scale and global practices. Metaphorically, practically, materially, the collapsing of the natural and the machinic was carried out on the largest scale. The extensive network of listening stations that intercepted enemy short-wave messages, passing them on to signals intelligence for enemy location, traffic analysis, and code breaking, relied on these practices to monitor likely frequencies, detect messages, and pinpoint their origins.[19] Driven by this electronic espionage and their own communications needs, the Allies created a group of "prediction laboratories" to coordinate the enormous effort to map the global ionosphere using ionograms, to forecast its behavior, and to identify usable communication frequencies for operations anywhere in the world.[20] Located at Washington, London, and Sydney, the laboratories drew ionograms from a sprawling infrastructure of local field stations strategically scattered across the theaters of war, spreading across six continents and anchoring a global network of circulating personnel, instruments, and records.[21] (See figure 1.2.) By 1944, the work of the prediction laboratories had carved the globe into a vast machinic order of three broad communications zones—Polar, Temperate, and Equatorial—each charted and defined by the reliability of radio receptions. Polar communications were the most heavily disrupted; Temperate were the most reliable, with Equatorial lying in between.[22] The zones, originally defined by technological effects, now came to define corresponding "ionospheres"—polar, temperate, and equatorial—each characterized by specific natural phenomena believed to determine radio propagation. The model that emerged was of a spherically symmetrical ionosphere whose axis of symmetry coincided with Earth's rotational axis, making ionospheric conditions identical along a given parallel of latitude. In theory, then, measurements along a single line of longitude (the Prime Meridian, say) could be used to give radio-frequency predictions for any place in the world.

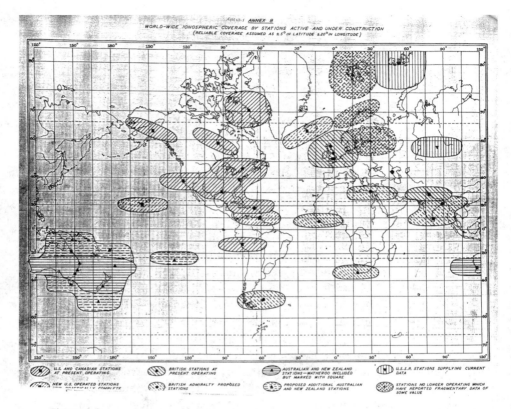

Figure 1.2

A map showing the locations of ionospheric stations in 1945. Since ionospheric conditions were believed to be constant along a given line of latitude, stations were arranged so that any point of interest on the globe could be connected along a parallel of latitude to the "shadow" of an ionospheric sounder, allowing predictions for usable radio frequencies at the original point. Three stations were in operation in Canada: one outside Ottawa, one at Churchill, and one (run by the United States) at Clyde River on Baffin Island. By 1947, Canada would take over the Clyde station and would establish two more stations on the east and west coasts—Torbay and Prince Rupert, respectively. Source: "The Status of the World-Picture of High Frequency Radio Propagation: Report by Working Committee to Consider CX/WP 29," Appendix B (1945), Library and Archives Canada, Record Group 24, volume 4058, file NSS 1078–13–8. © Government of Canada. Reproduced with the permission of Library and Archives Canada (2016).

Secret maps produced by the US Interservice Radio Propagation Laboratory (IRPL) revealed the model and the strategy behind it. As of March 1944, the Allies controlled forty ionospheric field stations either already operating or nearing completion—six each in Australia and New Zealand, twelve in Great Britain, twelve in the United States, two in Russia, and two in Canada.[23] Any point on the globe from Iceland to Tierra del Fuego could be connected along a parallel line of latitude to the "shadow" of an ionospheric sounder. Since ionospheric characteristics were believed to vary only with latitude, the data from the corresponding sounder could be used to generate predictions for usable radio frequencies at the original point of interest, wherever in the world it happened to be. (See figure 1.2.)

The Politics of Anomaly

In late 1943, a discovery in an unexpected place violently shook this neat global model of the ionosphere. Since about 1941, radio researchers, technicians, and telegraphists had observed that long distance radio circuits from India to England operated efficiently on frequencies higher than those forecast by the Allied prediction laboratories. To investigate those conditions, the British began operating a new ionospheric station in Delhi in late 1943, the only station covering the enormous stretch between the Middle East and Australia. Within a few months, ionograms from Delhi were suggesting something startling under the current model: that ionization above Indochina was more intense and supported communications on frequencies much higher than at San Juan Puerto Rico, which lay at the same latitude.

The set of conditions above Indochina fractured the tidy model of a spherically symmetrical ionosphere and the prediction systems based upon it. Quickly dubbed "the Longitude Effect," the discovery suggested that the morphology of the ionosphere escaped the neat latitudinal divisions of the geographic globe, varying instead according to both geographic latitude *and* longitude. In place of a sphere whose axis of symmetry coincided with Earth's rotational axis, geophysicists now envisioned a more complex orientation for the ionosphere with, at the very least, an axis of symmetry tilted in relation to the terrestrial globe.

The discovery had dire operational implications for the Allies. The uncoupling of geographic latitudes from constant ionospheric conditions suddenly pried open gaps in the "world coverage" provided by Allied field stations. Before 1943, "covering" the globe with ionosondes had meant simply blanketing the full range of latitudes. The meaning of coverage now changed. Local conditions could no longer be folded into ionospheric data taken at the same latitude half a world away. The most serious gaps appeared in the Pacific Northwest, in the area between San Francisco and Fairbanks. With the increasing importance of the Pacific War, one secret report explained, the Longitude Effect posed "a serious threat" to proposed future operations in the North Pacific.[24] After examining the coverage of existing field stations as well as those under construction, the report concluded that at least one more ionospheric station was needed between California and Alaska, with a preference for Western Canada or the Canadian Pacific coast.[25]

But researchers tracked more than operational implications across the Pacific from Indochina. They tracked *explanations* too. Sir Edward Appleton, Britain's foremost ionospheric physicist, speculated that the elevated ionization and lowered layer heights of the Longitude Effect might be explained by their conjugate—the known depressions of critical frequencies and raised layer heights that occurred in high-Northern latitudes near the auroral zone. One of Appleton's contributions before the war had been to work out the equations describing the differential refraction of radio waves by Earth's magnetic field. Now he noticed that the Delhi station recording the anomalous effects lay furthest from the northern auroral belt. Since magnetic activity was known to depress the critical frequencies and raise the height of certain ionospheric layers, he suspected that the answer to the Longitude Effect might rest with ionospheric investigations in high latitudes.[26] The discovery of the Longitude Effect therefore had at least two important consequences for wartime ionospheric research: its existence complicated the model of the global ionosphere; and its details focused intense interest on the Northern polar regions, linking them with the tropics and suggesting that the strange behavior of radio sets from South Asia to the North Atlantic had their physical origin in the high-Northern ionosphere.

That linking of tropics and high latitudes, particularly the North, was not new. Historically, both notions of tropicality and nordicity had emerged, not in opposition to one another, but in tandem, as a contrast to the perceived normality of the temperate regions.[27] If European encounters with the New World were shaped by the idea of the wondrous or the marvelous, the scientific engagement with both the North and the tropics was organized around ideas of nature at the extremes, particularly embodied in anomalies.[28] In commenting on the tropics in the late 1860s, for instance, Alexander von Humboldt wrote that "nature in these climes appears more active, more fruitful, we may even say more prodigal of life."[29] Other commentaries would cut in the opposite direction, characterizing the tropics through narratives of disease that gave tropical nature a "pathological potency that marked [it] out from milder, more temperate lands."[30] What linked the two narratives was a concern with aberration: tropical nature was more active, more fruitful, but also more disease-ridden and more deadly than in temperate regions. And it was precisely this intensity that gave the tropics a privileged epistemic position. "Nowhere," Humboldt wrote, "does [nature] more deeply impress us with a sense of her greatness, nowhere does she speak to us more forcibly."[31] That vision of nature at the extremes similarly structured Northern and polar research from the seventeenth century on, informing the idea of remote Northern regions as "laboratories" used by natural philosophers and later scientists to turn extreme geographical locations into central cultural positions.[32] Geophysically, then, the proper opposite of the polar regions was not the (also anomalous) tropics, but rather the comparatively well-behaved temperate latitudes that defined them both. That opposition—between the polar and the temperate—would form the basis of a politics of anomaly that sought to reimagine Canadian national identity after the war.

The Geophysics of Peculiarity

The high-Northern latitudes had been a focus of intense concern for the Allies since the war began. The Battle of the Atlantic, the longest continuous military campaign of World War II, had made communications in the region a major priority. In response to Britain's use of convoys in the North Atlantic, the German navy had begun using a highly centralized form of submarine warfare, based heavily on encrypted short-wave radio transmissions. The system allowed submarines to operate in more lethal "wolf packs" (*Rudel*) with tactical operations controlled directly from Germany's submarine command headquarters (*Befehlshaber der Unterseeboote*, abbreviated BdU) in occupied France. When a German submarine located a convoy, it would shadow it, sending out information by short-wave on the convoy's location, bearing, and speed. BdU would then direct other submarines to converge on the convoy, waiting until the full wolf pack was in position before giving the order to attack after nightfall. The results were devastating; Winston Churchill would claim that the threat they posed was the only thing that frightened him during the war.[33]

The British countered in at least two ways. Intelligence and cryptographic teams, including Alan Turing's at Bletchley Park, aimed to decode the content of the submarines' transmissions to discover their location and operations. The less celebrated work of HF/DF (high-frequency direction-finding) teams tried to pinpoint the origin of the signals.[34] In the vast stretches of the North Atlantic, however, turbulent ionospheric conditions generated by the auroral zone caused serious errors in the direction-finding bearings, helping U-boats to operate with impunity. After the Admiralty's direction-finding operations in London were bombed in early 1941, British officials sought to transfer high-frequency monitoring of the North-Western Atlantic to Canada and simultaneously urged the Canadians to use ionospheric data to improve its accuracy.

In May 1942, the Royal Canadian Navy responded by assembling a group of researchers housed in Section 6 of its Operational Intelligence Centre (OIC/6)—a unit concerned, like its British counterpart, with signals intelligence and anti-submarine warfare. In contrast to both the American and British operations, though, the smaller size of the Centre's nine sections made it possible to house their various activities—submarine tracking, high-frequency direction finding, and discrimination—in adjoining rooms, making the operation faster and more efficient.[35] Led by Frank Davies, a Welsh émigré attached to British intelligence early in the war and later seconded to the Royal Canadian Navy, OIC/6 carried out atmospheric investigations to improve the interception of German radio transmissions in the high-frequency range and to pinpoint the location of enemy submarines. With the situation in the North Atlantic now desperate, and with ships being sunk off the coast of Newfoundland, the Navy, anxious to secure any advantage it could, expanded the group's activities to include all questions concerning radio propagation in the vicinity of the northern auroral zone, including short-wave radio blackouts that characterized the region.[36] From two field stations, one outside Ottawa and the other on the shores of Hudson's Bay at Churchill, Manitoba, the team began detailing the structure and dynamics of the high-latitude ionosphere. Throughout the most intense phase of the Battle of the Atlantic in 1942, Davies' group worked furiously to produce ionograms for their own research as well as for the "centralizing laboratories" in Washington and London. As their research expanded the following year, the discovery of the Longitude Effect changed its status entirely.

Geophysical anomalies had always fascinated Frank Davies. Born in Wales, he had emigrated to Canada in 1925 and had worked as a guard on the Canadian Pacific Railway, escorting Chinese laborers to Vancouver, before taking up a position as Lecturer in Mathematics and Demonstrator in Physics at the University of Saskatchewan. After earning a master's degree in physics at McGill University, he served on a number of scientific expeditions to the Arctic and the Antarctic in the 1920s—as a physicist on Richard Byrd's first Antarctic expedition in 1928 investigating ice-crevasse temperatures, crystal formations, magnetic phenomena, and aurorae and receiving the Antarctic Medal from the US Congress for his efforts; as leader of the 1932 Canadian Second Polar Expedition, standardizing magnetometer readings and analyzing them for the Carnegie Institution in Washington after his return in 1934. Davies was fascinated by magnetic, auroral, and ionospheric observations. What linked his research across three continents and two poles was an intense interest in the geophysics and meteorology of extreme environments. Those interests would stay with him all his life. Long after his professional duties had turned administrative, careful meteorological notes on temperature, wind properties, and precipitation found their way into his diaries. After a posting as director of the Carnegie Geophysical Observatory in Huancayo, Peru in the late 1930s, he joined British Intelligence in the early years of World War II. He was seconded to the Royal Canadian Navy and then to Canada's National Research Council before Canadian officials asked him to head OIC/6.[37]

Preferring the term "longitude anomaly" to "effect," Davies seized on the discovery as an important impetus to his team's work and made the operational implications clear to his superiors. "It is probable," he wrote with his colleague Jack Meek, "that the whole area of war of the India-Burma-Malaya-China sectors needs frequency predictions materially different from areas in similar latitudes in America, Europe, and Africa." Davies went on to pair the effects over India with what had become the other archetypal anomaly: the high-latitude ionosphere, the only other communications zone that broke with the common understanding of ionospheric morphology. In a report prepared by Davies and Meek, the two noted that "the early, rather simple hypothesis that one could predict transmission frequencies accurately for the whole globe from measurements along a single longitude, was known to be inaccurate as far as high latitudes was concerned, but it is a surprising fact that this does not apply to low latitudes all around the globe." "There is, as yet," they concluded, "no explanation for this."[38]

Linking the tropical and the polar even more closely in the months that followed, Davies began seeing the high-latitude ionosphere as nothing less than a possible cause, rather than a mere analogue, for conditions over the tropics. The high-latitude ionosphere represented the beginning of a complex causal chain of geophysical effects that started over Canada and ended over Indochina, potentially explaining the Longitude Anomaly as a whole. Davies went further. With the high-latitude ionosphere now holding the promise of explaining radio disruptions from the North Atlantic to the Indian subcontinent, he began viewing the atmospheric region as a way to reposition not only his group but also his adoptive nation. Like many, Davies felt that Canada's wartime scientific efforts had developed under the shadow of the British and the Americans. The entire war had been characterized by that larger anxiety over status and influence, and it had spilled over into its most important scientific efforts, including nuclear research.[39] British and American work had dominated ionospheric physics since the interwar period, but that work had been focused mostly on the "temperate" ionosphere of the middle latitudes. High-latitude research would both distinguish Davies' group from other ionospheric teams and make use of Canada's northern position to set the nation apart from its allies.

The broad lines of Davies' approach were captured in a secret document produced in late 1944, "The Application of Ionospheric Measurements." The report aimed to give military officials a general overview of the role of ionospheric data in securing reliable short-wave radio. Less than a third of the way through the document, after brief discussions of wave motion and ionospheric propagation, Davies and his colleagues presented their readers with a view of the atmosphere in cross-section. (See figure 1.3.) Starting at the surface and moving up through the troposphere and stratosphere, the document sketched out the various ionospheric layers—the D, E, and F regions. In the midst of the electron plasma, Davies and his group situated two important geophysical phenomena: cosmic rays, represented as particles showering through the upper ionosphere; and the aurora borealis, the northern lights, believed to be associated with (if not responsible for) radio blackouts and short-wave disruptions throughout the Northern Hemisphere.[40]

Within the context of wartime ionospheric investigations, this was no generic trans-local representation of atmospheric structure. Rather, it was part of a still speculative and highly circumscribed *natural order* of high-Northern latitudes—a physical and conceptual ordering of geophysical phenomena that decades of geophysical studies had implicated in ionospheric structure and dynamics, and specifically in radio propagation near the auroral zone. Weaving back and forth between this speculative geophysics and its likely effects on radio propagation, Davies and his colleagues arrived at a corresponding *machinic order*—a spatial ordering of short-wave radio behavior that pointed to communications difficulties in the North Atlantic and in South Asia. (See figure 1.4.) That machinic order was also highly speculative, based, as it was, on the behavior of hypothetical machines located throughout the country. Over the coming two decades, through the detailed mapping of radio facilities and reception conditions, it would increasingly be fleshed out with actual machines and their behaviors, culminating in the radio geographies of the late 1950s discussed in chapter 7.

Having established the major issues of wartime short-wave communications as a problem of the high-latitude ionosphere, Davies and his group

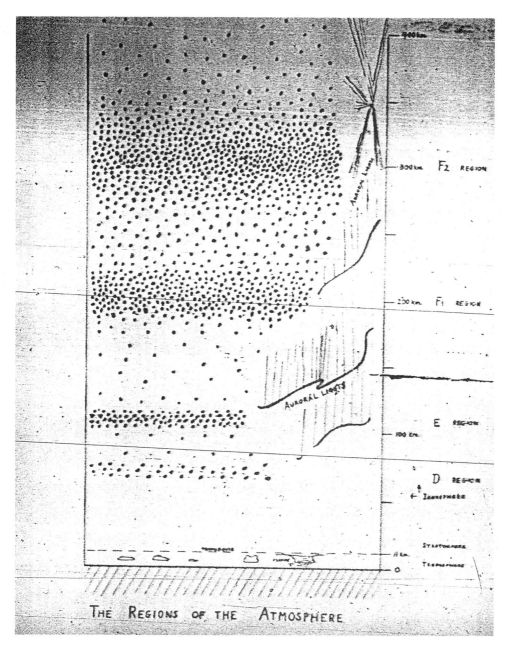

Figure 1.3

A 1944 illustration of regions of the ionosphere (D, E, F1, F2), showing cosmic rays (shown as the shower emanating from upper right), and the Aurora Borealis (shown throughout the ionospheric regions). Source: Canadian Radio Wave Propagation Committee, "The Application of Ionospheric Measurements" (1944), Library and Archives Canada, Record Group 24-C-1, file 9147–2-1, p. 13. © Government of Canada. Reproduced with the permission of Library and Archives Canada (2016).

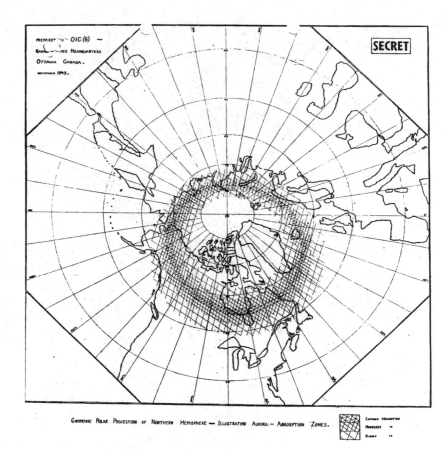

Figure 1.4
Radio absorption zones of the Northern Hemisphere. This map, drawn by members
of OIC/6 in November 1943, illustrates the spatial extent of short-wave radio disrup-
tions, which coincides roughly with the appearance of visual aurora. Note its dis-
proportionate extension into Canada. Source: Canadian Radio Wave Propagation
Committee, "The Application of Ionospheric Measurements" (1944), Library and
Archives Canada, Record Group 24-C-1, file 9147–2-1, p. 30. © Government of Canada.
Reproduced with the permission of Library and Archives Canada (2016).

now began recasting that problematic relationship, which achieved its most distilled and potent form above Canada. In its 36 pages, the report emphasized two points that Davies would soon place at the center of arguments for continuing his group's work after the war. The first was that radio disruptions in northern regions were linked to unique and *characteristically* Northern geophysical phenomena (aurorae, magnetic storms, ionospheric turbulence, etc.); the second was that Canada, because of its Northern character (as *defined* by these geophysical phenomena), was in a unique position to investigate those phenomena and solve short-wave radio problems throughout the world.[41] Echoing the language of both arctic and tropical epistemologies, Davies presented Canada as a "natural geophysical laboratory" in which to investigate the high-latitude ionosphere and to trace out the precise links between Northern geophysical phenomena and unreliable radio.[42] In this way, the document pointed beyond the immediate narrow technical concerns of the wartime North Atlantic to the national and international importance Davies would claim for his group's research after the war.

Northern Communications

By the time OIC/6 issued its report, the idea of Northern communications in Canada resonated deeply with two elements of national mythology. The first was what Shelagh Grant has described as the "core myth"—the idea that Canada was an essentially Northern nation whose distinguishing characteristics—the winter, the tundra, the northern lights—granted it a unique character and identity.[43] The "core myth" entered World War II bolstered by a century of geophysical investigations that placed the North Magnetic Pole in Canadian territory and gave Canada the largest portion of the aurora borealis of any country in the world. The second belief involved a parallel "communications myth" based on the conviction (which went back to the days of the transcontinental railway in the late nineteenth century) that Canada was the first and perhaps the only communications state—a polity owing its very existence to the establishment of reliable communications.[44] Together, the two myths underpinned a belief that the inherent properties of the North thwarted communications and that the unique struggle against these obstacles served to define the nation.

Those associations drew additional force from an immediate political crisis. In 1943, in a breach of protocol that he justified by its urgency, the British High Commissioner to Canada, Malcolm MacDonald, wrote a personal note to the Canadian War Committee. MacDonald had been invited by Canadian officials to tour the joint defense projects built by US forces in the Canadian Northwest between 1942 and 1943—projects that included thousands of kilometers of highway to defend against Japanese invasion of Alaska ("the Achilles heel of American defense") and a string of airfields to supply Allied forces in Europe.[45] MacDonald worried that fears of Japanese invasion were a ploy on the part of the United States. Despite formal agreements, the US was positioning itself to gain complete control of air routes to Europe and Asia after the war. Infrastructure projects were simply "one of the fingers of the hand which America is placing more or less over the whole Western Hemisphere."[46] After his visit in March 1943, MacDonald warned that Canadian officials in the Yukon and in British Columbia seemed unable to control day-to-day events, US Army personnel were openly referring to themselves as an "Army of Occupation," and a second undisclosed chain of airfields was being built. MacDonald claimed to have confidential information from US Army sources that the roads and airfields were being planned with an eye to commercial aviation and transport after the war and for "waging war against the Russians in the next world crisis."[47]

MacDonald's memo landed on the desks of a small but influential group of former academics known in Canadian government circles simply as "the intelligentsia." Heavily concentrated in the Department of External Affairs, and centralist and interventionist in their outlook, they had observed at first hand what they viewed as the neo-imperialism of their allies.[48] After the fall of France in June 1940, London's War Rooms had seemed to "automatically" determine Canadian policy, and after the attack on Pearl Harbor 18 months later, the United States had increasingly treated Canada as "an internal domestic relationship rather than an international one."[49] Locating the problem in weak and reluctant diplomacy ("the strong glove over the velvet hand"), members of the Department of External Affairs began exercising a new muscularity in Canadian foreign policy. The ubiquitous Hugh Keenleyside, a former history professor at Brown, Penn State, and Syracuse, recognized that a more assertive foreign policy would require a break in Canada's historical understanding of itself. Keenleyside argued that Canada could only gain autonomy and status by severing its long-standing dependence on Britain and its emerging dependence on the United States. One of his most imaginative colleagues, Escott Reid, put it this way: "If we don't want to be a colony of the United States we had better stop being a colony of Britain."[50]

Inspired by MacDonald's report and recognizing the growing geopolitical importance of the Canadian North, Keenleyside and his colleagues seized on the region as the site to begin breaking Canada's historical dependency. Drawing on the long-standing visions that turned Canada's supposedly

Conclusion

World War II produced a succession of natures imagined, articulated, and shaped by the behavior of wartime technologies. But few had a wider global reach than the ones produced by ionospheric research. Under the demands of tactical communications, intelligence operations, and anti-submarine warfare, researchers increasingly cast the ionosphere not as part of a coherent natural order but rather as part of a global communications system. They collapsed the lengthy chains of techniques and calculations that had linked geophysics to radio propagation and nature to technology early in the century. Providing indispensable support for the Sciences of the Enemy, they operationalized the ionosphere, creating natural-machinic hybrids that reinterpreted ionospheric phenomena and atmospheric morphology through the sometimes erratic behavior of wartime machines. Through practices centered on and designed to link the local and the global, they ultimately produced a fusion of nature and technology that dominated wartime work, and that, in a highly specific form, became central in attempts to refashion at least one postwar nation.

If that association began as a general link between geophysics and machinic behavior, it quickly became highly specific. The discovery of the Longitude Anomaly broke with the neat, efficient model of the ionosphere and its associated geography of radio effects, opening up dangerous gaps in the coverage of the ionospheric network that had been established during World War II and suggesting a possible connection between radio problems in Indochina and phenomena in the high-latitude ionosphere above Canada. Linking the natural phenomena of the tropics to the polar regions, and setting them against the well-behaved temperate ionosphere, Davies and his group suggested that the Northern high-latitude ionosphere might be used not only to fill gaps in ionospheric coverage in the Pacific Northwest but also to overcome problems affecting short-wave radio reception around the world. In the 1970s and the 1980s, climate scientists would again make similar claims for the role of the North in global knowledge.[63] But Davies' group, immersed in problems of direction-finding and anti-submarine operations, translated this generalized relationship into a specific connection between the ionosphere above Canada and the unreliable operation of radio sets near the equator.

As World War II drew to a close, Davies and his colleagues turned their investigations into arguments for an archetypical *Canadian* geophysical order and its relationship to failing machines, weaving it into the much broader national mythology surrounding the North and communications and into the immediate attempts to elevate and reimagine Canada at the end of the war. That overall trajectory—from global to polar to Canadian—forged a relationship between nature and technology whereby the malfunctioning of machines increasingly said something about turbulent nature, technological failure, and national identity. Through the work of Davies' group, radio blackouts and disruptions became machinic manifestations of a specific and increasingly characteristic Canadian natural order—so much so that this relationship between nature and machines would eventually be used to characterize the nation far beyond the walls of the laboratory. The immediate challenges the members of Davies' group faced were how to detail that natural order and how to make it and its characteristic phenomena authoritative. In tackling those challenges, they would have to transform two other sites of unreliability: the production of ionograms and the interpretation practices that made them legible. Their efforts to weave together the natural and technological orders through those privileged records are the subject of the next two chapters.

2 Machines and Media

What did it take to link machinic and natural orders in the way Davies and his group proposed? Throughout World War II, it had often been difficult to determine whether humans, machines, or natural phenomena were to blame when radios failed. Barring complete blackouts, the static from auroral displays or tropical thunderstorms often mimicked the effects of inexperienced operators or faulty equipment. That same uncertainty beset the operation of ionosondes—the instruments that produced the all-important ionogram, and that therefore mediated between natural phenomena and machinic effects. Were the peculiar traces on the scientific records evidence of distinctive phenomena above Canada, or artifacts created by unreliable personnel and malfunctioning instruments? In order for Davies' group to begin tracing out the links between distinctive natural phenomena and disproportionately failing machines, they needed to make the scattered Canadian field stations function as a unified whole.

That challenge was particularly daunting as World War II drew to a close. While the network of stations expanded after the discovery of the Longitude Anomaly in early 1943, its corps of experienced personnel had begun to fracture seriously. Each station was now run by a different branch of the Canadian military, with Frank Davies' group simply providing oversight and training. Demobilization in the late summer of 1945 made the situation more dire by generating a steady stream of personnel that changed every few months, particularly in the remote northern locations. The improvised ionosondes assembled from spare parts during the war also began breaking down or malfunctioning, requiring extra care and expertise to keep them running. Davies' group ultimately lost control over how ionograms were being generated at the remote sites. In doing so, they also lost control over whether the ionsondes could faithfully capture complex high-latitude phenomena, and whether the phenomena they did capture could be trusted, circulated internationally, and assembled into reliable representations of the upper atmosphere above Canada.[1]

In the late summer of 1945, Davies therefore turned to the possibility of standardized ionosondes as a way of resolving doubts about the performance of both humans and instruments.[2] Within the historiography of science and technology, standardization has generally been seen as a solution to problems of replication. It is a historical development—made possible by precision manufacturing and the rise of metrology—that makes it possible for scientific results to travel.[3] Davies' deliberations over standardized instruments suggest the mistake of seeing standardization solely as a drive toward sweeping uniformity. For Davies, standardization supported two kinds of identity. It would make specific functional characteristics of the machines identical across North American and British field stations; but in doing so it would also help establish claims about the legitimate difference of high-latitude phenomena, establishing the accepted and authoritative background against which meaningful distinctions about the Canadian ionosphere could emerge. In pushing for a standardized instrument across the laboratories and field stations of the former wartime Allies, Davies was concerned about the parity of scientific data; but he also saw how standardized machines might generate a uniform field against which the distinctive natural order that characterized Canada and its communications disruptions would come into stark relief.

What it took to establish those connections is the subject of two chapters. Chapter 3 explores the new reading regimes DRTE staff developed to make high-latitude ionospheric phenomena legible and to combine them with other scientific records into a visualization of the geophysics responsible for radio failures. This chapter sets the stage for those developments by dealing with a different machinic order—one centered on the behavior of scientific instruments rather than on short-wave radios. It focuses in particular on the attempts to create an instrumental collective out of the Canadian field stations: not a system or a network, but a collection of individual ionosondes collectively responsible for mapping a particularly important sector of the global ionosphere. In doing so, it examines how the efforts to standardize these instruments opened up important possibilities for variation and difference in international collaborations. Those efforts formed part of a more general trend in postwar radio engineering, meteorology, and geomagnetism that aimed at both sharing information and producing a shared understanding of the world as a whole.[4] But for Davies and his group, the new ionosondes were a way of solving all-too-local problems of the place of experiment, the people occupying it, and the larger machinic and natural orders that depended on them. Standardizing machines, they hoped, would make difference possible by first making things the same.[5]

Ionograms and the Place of Experiment

Immediately after World War II, the sprawling networks that had been set up to collect high-Northern ionospheric data for wartime operations began to unravel as Soviet stations stopped sending their ionospheric data to the West. At the same time, those data became even more important for establishing reliable long-distance communications in the polar regions and, in particular, for securing the North American Arctic against Soviet invasion or attack. The accuracy of frequency predictions for high-Northern latitudes therefore came to depend almost exclusively on Canadian data (with the exception of the US station at Barrow, Alaska). After the discovery of the Longitude Anomaly, officials had proposed adding additional ionospheric field stations to the existing Canadian network. Frank Davies had acted quickly on those recommendations, shepherding the requests for two new stations through official channels. Scarcely two weeks later, Davies had laid out his vision for the new field stations, a plan dedicated to the continuous production of ionograms and organized geographically around the auroral zone.

The two new stations—one for Prince Rupert, British Columbia, and one for Torbay, Newfoundland—were expected to begin operating in June 1945. The Department of Transport would take over operation of the station at Clyde, Baffin Island, which had been operated by the Carnegie Institution since September 1943. A longer-term program would set up a perimeter of additional stations along the southern extent of the auroral zone; a series of mobile observatories—using ships, planes, trucks, and even a converted railway car—would slice across the zone; the station at Churchill, Manitoba would provide data from within the zone itself.[6]

The plan was to move in three phases toward a dedicated prediction service for Canada. First, the field stations would record and analyze greater quantities of ionospheric data over Canada; once enough data had been collected and enough staff had been assigned, the lab would move to a second stage involving Storm Warnings, bulletins alerting radio operators of impending communications blackouts; the third stage would see a dedicated prediction service for Canadian short-wave communications.[7] This third stage depended on suitable methods for graphing high-latitude phenomena and for analyzing data from the field stations.[8]

For Davies, who had been brought up in the tradition of geophysical field investigations, the field stations were the heart of ionospheric research. Remembering OIC/6's wartime investigations, he would recount the long hours of tedious desk work in Ottawa "lightened by visits to the ionospheric stations."[9] He now styled the new field stations after the isolated and bucolic outposts he knew so well. Each station would cost about $4,500. It would include living quarters for the five-man staff (exclusively male) and separate rooms for ionospheric equipment, repair services, and data analysis and reduction. Davies' group was officially in charge of directing the operation of the stations and of collecting and coordinating their data. But the enormous distances between stations made direct supervision impossible. As a result, Davies favored making each station self-sufficient, capable of producing data without interruption and forming the focus of the professional and social lives of the operators who ran it. Housed together, working in shifts, and sharing an "intelligent understanding" of ionospheric observations, the operators would form dedicated and cohesive scientific units scattered around the country and furnishing data on the peculiar ionospheric conditions across the auroral zone.[10]

Life at the stations revolved around the ionosonde, an instrument that had grown out of the same practical and material culture as radar. Because of wartime equipment shortages, Canadian ionosondes of the late 1940s were often assembled from surplus radar equipment. The common instrument at the Canadian stations was the manual ionosonde, which produced what were known as A-scans (panel a in figure 2.1).[11] To construct a height-versus-frequency graph, or ionogram (panel b in figure 2.1), operators of field stations measured certain characteristics of these radio returns (either directly on the oscilloscope or on photographic records) while varying the frequency of the instrument.[12] The automatic ionosonde automated this process by displaying a so-called B-scan—an image in which the frequencies of radio echoes blanked out a corresponding portion of the sweep trace of the oscilloscope proportional to the transit time of the pulse (panel c in figure 2.1). Focusing the resulting image on 35-millimeter film while moving the film at constant speed created the same h'f curves without the difficulties and uncertainties of manual intervention (see panel d in figure 2.1). In early models, a camera unit attached to one of the oscilloscopes was triggered through a foot pedal by operators monitoring the second display. Later modifications would eliminate the monitoring tube and attach an automatic 16-millimeter motion-picture camera. Operated through a series of gears, the camera would parse the play of light on the oscilloscope into successive "snapshots," each frame functioning as an individual ionogram.

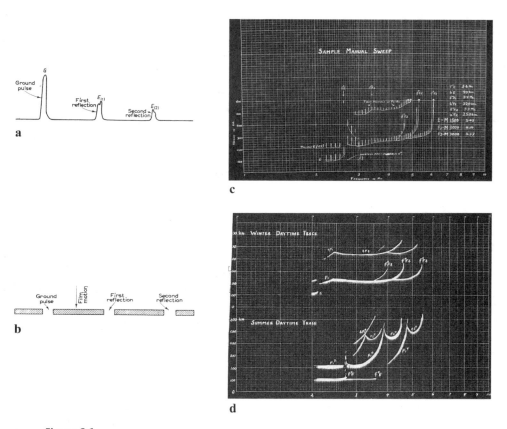

Figure 2.1

The manual ionosonde produced an A-scan (a), which was displayed on the instrument's cathode-ray tube. Technicians measured and plotted the height of the pips to produce a manual ionogram (b). The automatic ionosonde automated the process by blanking out portions of the trace to produce a B-scan (c) and projecting them onto moving film (d). Sources: "IGY Instruction Manual, the Ionosphere: Part I, Ionospheric Vertical Soundings," ed. W. J. G. Beynon and G. M. Brown, *Annals of the IGY III* (1958): 32 (reprinted with permission of Elsevier); Canadian Radio Wave Propagation Committee, "Instructions for Observers: Canadian Ionospheric Stations," July 1944, Library and Archives Canada, Record Group 24, file 9147–2; © Government of Canada (reproduced with permission of Library and Archives Canada).

Davies' plan was to equip the stations first with manual, direct-reading instruments like the one used at the naval ionospheric station at Chelsea, just outside Ottawa. Simple and reliable, that instrument had given what Davies described as "very efficient service."[13] Eventually, the manual equipment would be replaced by automatic recorders designed and built by the National Research Council; the older equipment would be used as backup and for absorption measurements. Davies' initial intention in the late spring of 1944, when details for the new stations were being finalized, was that the automatic equipment would establish "parity" between the Canadian stations and the international program (particularly the work being done at the National Bureau of Standards, where the automatic ionosonde had been invented).

As his group moved toward the creation of the Radio Propagation Laboratory, Davies began to see the automatic equipment as a solution to a loss of control over the field stations and the problems of experience and training that came with it. The ionosonde had long been located at the intersection between the field and the laboratory, between observation and experiment. The automatic version of the sounder had been created in response to deep concerns about human and material resources, labor, and efficiency.[14] Its inventor, Theodore Gilliland, worried in the early 1930s that existing recorders would require hundreds of feet of film and half a dozen workers to complete a 24-hour run of observations. As Frank Davies and his colleagues became increasingly worried about the work being done at the stations, the ionosonde began to figure as a technical solution to a human problem: how to ensure that the effects displayed by the machines reflected natural phenomena rather than human artifacts.[15]

The Machine in the Wilderness

The station at Prince Rupert illustrated these anxieties over people and machines. Incorporated in 1910, Prince Rupert had been British Columbia's first planned city. Located along the northwestern shore of Kaien Island, at the mouth of the Skeena River and about 770 kilometers north of Vancouver, it lay on the deepest natural ice-free harbor in North America. Charles Melville Hays, a former railway manager who would drown on the *Titanic* in 1912, had conceived of it as a rival port to Vancouver and a perfect terminus for maritime trade and rail travel. After 1910, the city had first served as a tent camp and a western terminus for the Grand Trunk Pacific Railway. By the 1930s, Prince Rupert was a major seaport. After the attack on Pearl

Figure 2.2
The ionospheric field station at Prince Rupert, British Columbia. Field stations often resembled modest cabins, and were located in isolated areas (as above) to shield them from radio interference and to protect the confidentiality of their data. Source: Letter from Lt. Robbins to Maj. MacLeod, January 2, 1946, Library and Archives Canada, Record Group 24-C-1, file 9147–2-1, microfilm reel C-8437.

Harbor in 1941, the US government used it as a point of embarkation for troops and munitions headed to the Aleutians and the Pacific. The town was fortified with gun emplacements and personnel forts; a submarine net closed off the mouth of the harbor; navy gunboats patrolled the coastline. The population tripled.[16] A number of Canadian troops were also posted there, including the Royal Canadian Corps of Signals, whose barracks, workshops, and offices were built adjacent to the Army's Signals group at the end of Second Avenue and Seventeenth Street.[17] To take advantage of the local material and human resources, the ionospheric field station was set up in the complex of buildings housing the Royal Canadian Corps of Signals, amid the sprawling antenna arrays of the transmitting building behind the Army barracks. (See figure 2.2.)

Technical plans designating the new building as Hut #10 were drawn up in March of 1945. The building itself resembled a moderately sized cabin. In accordance with the planners' instruction, all the lights featured pull-chain switches. Initial plans called for oil-burning stoves to keep the main spaces livable in winter. Although manual ionospheric equipment required only a workshop and an operating room, a darkroom was added for the film development that would eventually be required for the automatic sounders.

The process that turned radio pulses into the raw material for transmission predictions and ionospheric morphology was captured in the architecture of Hut #10. The building included four main sections. A small room (illustrated at upper left in figure 2.3) was set aside as a workshop for repairs to the ionosonde and to peripheral equipment such as antennas and power supplies. An "Operating Room" held the main ionosonde unit. Electrical leads connected the device to the enormous kite-shaped rhombic antennas mounted outside. It was in the Operating Room that operators drawn from the Army's signals division initiated sweeps of the ionosphere at least once an hour, and sometimes as often as four times an hour, 24 hours a day. The operators of the manual ionosonde would monitor the small cathode-ray tube of the recorder, setting the receiver to the lowest limit of the frequency range (2 megacycles) and then tuning the transmitter to the same frequency; beginning the slow and detailed process of measuring the width of the pips at their half-height, plotting a line equal to the height variation on logarithmic paper, then varying the frequency and beginning the measurement process again. The entire frequency sweep and the creation of a single graph would take about 15 minutes.

Once the graphs were complete, operators transferred the record to the "Office Proper," where clerks carried out careful scaling measurements on the curves, determining layer heights and critical frequencies and laying rulers, sliders, and transparencies over the graphs to produce transmission frequencies and skip distances for radio. The data would then make their way to Davies' group in Ottawa, where they would be combined with traffic analyses and with auroral and atmospheric data and then passed on to Washington and London. With Soviet data arriving either late or not at all, the value of frequency predictions for Canada and adjacent areas, including the North Atlantic and parts of the Eastern Hemisphere, depended almost entirely on what occurred in these rooms at the five Canadian stations.[18] These two sites—the Operating Room, with its emphasis on faithful recording, and the Office Proper, with its emphasis on accurate reading and interpretation—were the foundational spaces for the work of the field

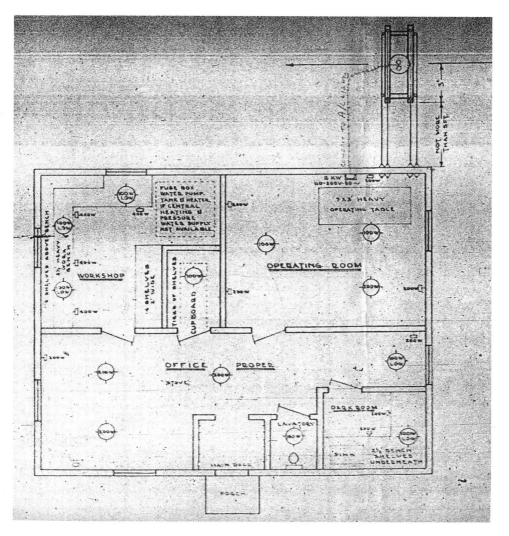

Figure 2.3
The floor plan for the Prince Rupert Ionospheric Station. Stations were generally divided into four rooms: a workshop, an operating room, a darkroom, and an "office proper." The introduction of the panoramic ionogram was meant to eliminate possibly unreliable practices in the operating room. Source: Lt. Robbins, "Recommended Plan for Ionospheric Station, Prince Rupert, B.C.," March 3, 1945, Directorate of Signals "Army," Library and Archives Canada, Record Group 24-C-1, file 9147–2-1, microfilm reel C-8437.

stations and for Davies' emerging program of research on the high-latitude ionosphere.

In the summer of 1944, the Scientific Staff began directing enormous efforts at what went on in these rooms. After World War II, Davies' group had begun shifting its emphasis and methods. In contrast to the continued drive toward correlating atmospheric data with radio behavior that emerged from the war, Davies' group had begun to distinguish itself by emphasizing the physical and the causal rather than the statistical and the correlative.[19] The move mirrored changes in other areas of meteorology, specifically the early-twentieth-century shift in weather forecasting away from statistical patterns and toward physical or dynamical models.[20] Although bulk ionospheric data were still important for frequency prediction, the Canadian group emphasized the ability to capture and identify the distinctive characteristics of the high-latitude ionosphere—what Davies had described as the "new or unusual phenomena" that would drive advances in the field and would characterize Canadian activities.

This focus on novel phenomena that generated both communications failures and anomalous ionograms underlined the need to discriminate between genuine geophysical phenomena and human-induced, "machine-made" artifacts. The practices that produced ionograms had evolved as a mutual imitation between humans and machines. On manual sounders, operators were required to measure pips "in order to produce a record somewhat similar to the [automatic] photographic equipment"; the automatic recorder, for its part, was arranged "so that the pips for all frequencies are recorded *as they would be plotted using a manual equipment.*" Automatic machines were designed to trace ionograms like skilled workers; workers learning the skill of tracing ionograms would perform (by hand) the actions carried out by ideal machines. As long as the ionosphere was settled, operators could use rules based on idealized "typical" ionograms to guide their work. Where the reflections were more complex or changed rapidly, as they did at high latitudes, operators had to rely on their own discretion. As the instruction manual for the staff explained, "In the case of fluctuating pips or those with more than one peak, the interpretation is left to the judgment of the observer."[21]

Fluctuating pips were precisely what interested Davies and his group now; problems of judgment were exactly what worried them. The rapidly changing pattern on the oscilloscope display characterized the turbulent ionosphere above Canada. Recording those pips faithfully did not require highly skilled personnel, according to Davies, but it did require sound judgment. The most obvious choice to carry out the work were wireless telegraphists from the military's radio wave propagation groups. They all had experience in radio propagation. Many had worked in direction finding, and distinguishing between legitimate targets and radio artifacts had honed their ability to make careful technical judgments. As one source explained: "Being on his own when a Bearing [sic] is intercepted, an operator must have sufficient judgment to report on the reliability of the bearing as well as identifying the type of signal. No one else in the ship can do this and the captain of the ship or the escort has to decide on the action to be taken on the verdict given by the operator."[22]

Like reliable direction finding, in which ionospheric data often informed judgments about targets and bearings, sound ionospheric observations also rested on a combination of moral disposition and experience. "Personnel selected for the Ionospheric stations should be intelligent and keen on their work," Davies explained. "They should be experienced with ionospheric propagation so as to see the crucial application of their work to communication problems throughout the region."[23] Davies had taken pains to explain the production and interpretation of ionograms in terms of radio propagation, expertise that telegraphists and wireless operators—the ideal staff for the stations—would possess. But the use of military personnel was only provisional. According to Davies' plan, the stations would eventually draw their personnel exclusively from the Radio Division of the National Research Council. The NRC, the only organization with extensive prewar experience in ionospheric research, had been the central site of Canadian wartime signals and radar work. During the war, researchers from its Radio Division had designed and built the ionosondes that were used by Davies' group at Chelsea and at Churchill. Bolstered by visits to RPL to reinforce their interests in using ionospheric data, the NRC staff would (Davies hoped) combine the character and experience necessary to make their work dependable.[24]

The perceived effects of the North on human operators, however, called their potential judgment into question. The North had quickly become a site of intensive research into what many considered to be unique problems in logistics, troop movement, and survival in extreme conditions.[25] The Defence Research Northern Laboratory at Churchill, where members of the DRTE staff were often stationed in the late 1940s, focused on the physiological and psychological effects of Northern work, including concerns about kinesthetic sensitivity, dexterity, vigilance, and the physical and cognitive abilities needed to maintain and operate machines.[26] In a secret postwar report on defense research in the Canadian Arctic, the geologist J. Tuzo Wilson, who served on the Arctic Circle group with Frank Davies, laid out a number of the factors affecting what he considered "the very serious problem" of suitable personnel in the North:

(1) isolation, loneliness and a severe sense of being cut off from civilization
(2) the depressing effects of infertile surroundings
(3) monotony
(4) unnatural seasons disturbing sleep or causing enforced idleness and boredom
(5) lack of internal resources and lack of incentive or enthusiasm when faced with the Arctic
(6) carelessness arising from being "bushed"—a state that arises due to some of the above factors even in much less severe conditions.[27]

Those effects combined with mass demobilizations to make experienced wireless operators and telegraphists, the usual operators of the stations, worryingly scarce.[28] In early May 1945, in a move that would transform the stations and ultimately drive the program for standardized equipment, NRC officials notified Davies and his colleagues that the NRC would not be able to operate the field stations after all, citing a shortage of staff. Instead, responsibility would be spread across the organizations that used ionospheric data—the Army would run Prince Rupert, the Air Force would operate the station at Torbay, and the remaining stations would be split between RPL, the Department of Transport, and the Navy. A "competent technical specialist, either civilian or service officer" would supervise the operations, aided by four assistants, one of whom would be qualified to maintain equipment.[29]

The havoc of demobilization severely disrupted the operation of the stations. Davies remarked in 1947 that the majority of personnel at the stations had been changed within a few months after being trained at RPL. Emphasizing the magnitude of the turnover, he counted seventy operators passing through three stations in less than two years. Training now had to be carried out on site, "sometimes by men who have not had sufficient training themselves." "Inaccuracy and serious errors in data have occurred because of this," he noted.[30] Initially buried in the details of confidential memos, the nominal loss of control over personnel at the stations spilled into view as the Armed Services redirected members of the already-thin technical staff to more pressing projects, constantly pulling from the stations the personnel who had not left on their own after the war.

The station at Torbay illustrated the loss of control. A Secret Organization Order drafted in October 1944 created the Torbay station as "an ancillary non-flying unit" of the Royal Canadian Air Force and placed it under the command of a Flight Lieutenant drawn from the Signals Division of the Air Force.[31] Nowhere in this formative document was Frank Davies or his group mentioned. The "assistants" of Davies' original memo, the order stated, would be a warrant officer and a flight sergeant (who together would act as Wireless Mechanics), a flight sergeant, and three sergeants (who would fill the roles of Wireless Operators, Ground). "This," the order explained, "is considered . . . the minimum [staff] capable of operating an Ionospheric Recorder successfully."[32] In early April, Air Force technicians acting under the station's commanding officer modified the manual iono-spheric equipment, adding a second three-inch cathode-ray tube with a rigid photographic attachment that allowed the operator to photograph the display. (See figure 2.4.) The camera could be actuated by a foot pedal or set to take pictures automatically at intervals from two seconds to one min-ute, recording the information on a continuous roll of 35-millimeter film. The film strips could then be projected onto a 6-by-8-inch or a 6-by-8-foot screen with a calibrating scale for analysis by the Signals Propagation Section of the Air Force.[33]

Through the photographic attachment, the modified equipment embod-ied a vision of ionospheric work that split the work of operating the iono-sonde from the work of making judgments about ionograms, tasks that Davies had necessarily united in the person of the operator. On May 10, a little more than a week after the station began operations, the RCAF drasti-cally revised its understanding of the work of No. 1 IRU, dividing operation from analysis in this way. "Techniques," the memorandum explained, "are similar to *Radar* in that continuous visual scanning of a small screen is required. WOG's [Wireless Operators, Ground] who are now being used, *are not trained for this work, and indeed may be considered misemployed.* Sufficient W/T [Wireless Telegraph] work exists for one WOG only."[34] The report went on to explain that the work was very exacting, "requiring continuous visual concentration on small moving images, easily producing severe eye strain."[35]

On the RCAF's view, *operating* the sounder had little to do with knowl-edge of wireless communications. Ionosondes were essentially radar sets, and should be staffed accordingly. Nor, on the RCAF's view, did the clerical work of the station have anything to do with the operation of the sounder, as they understood it. Scrapping the original complement of staff with direct radio experience, the memo recommended six "Radar Operators, Ground" (one flight sergeant and five sergeants) to take over the work of the ionosonde. All but one of the wireless operators would be eliminated (at the same time eliminating any members of the staff who had extensive training in radio transmissions). A Diesel Fitter would be brought on to

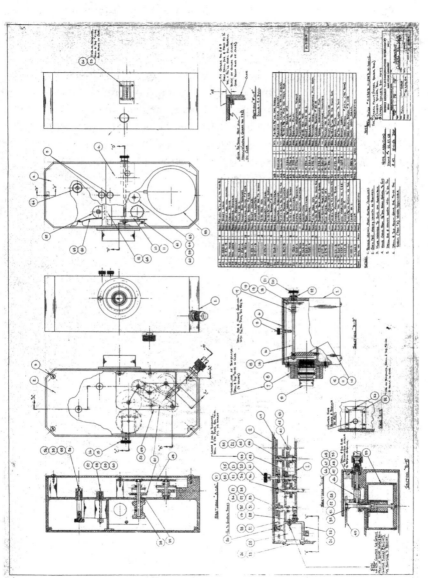

Figure 2.4

The LG17 photographic attachment. This technical drawing shows the camera installed as part of the ionosonde equipment at Torbay, Newfoundland. The camera was actuated by a foot pedal or set to take pictures automatically at intervals from two seconds to one minute, recording the information on a continuous roll of 35-millimeter film. Source: "Assembly Camera for LG17," Library and Archives Canada/Department of Communication fonds/

maintain the generator for the station, and a Clerk General would be appointed to deal with technical clerical work. The total number of staff members would now be twelve.[36] Experience in high-frequency radio propagation, which Davies had insisted throughout the war was central to the discretion of the operators and to "intelligent understanding," was put aside in favor of radar experience. The handbook vision of operators carefully analyzing the flickering, mutating images of the cathode-ray tube gave way to the view of radar operators mechanically triggering 35-millimeter cameras, when they were involved with the recording process at all.

The problem of personnel proliferated through a farrago of machines. Logistical difficulties and wartime scarcity of electronic parts made the reliability of ionosondes a major concern by the end of the war. One official report noted that "equipment in use at Canadian stations is obsolete, of a mock-up pattern, varies between stations, and is manually operated at three of the five stations."[37] "Without exception," Davies would explain some months later, "the recorders in use are wearing out. All were built hurriedly to fill immediate wartime requirements."[38] As with personnel, the conditions of the North were believed to make matters worse. The ionospheric station at Clyde, on Baffin Island, could be reached only once a year by boat, making the timely delivery of spare parts and electronic supplies difficult.[39] The modifications performed at Torbay were part of a wider culture of improvisation that had characterized wartime technical work, including the construction of ionosondes. On examining the manual ionosonde on loan from the Department of Terrestrial Magnetism, Torbay technicians fired off a terse memo to Davies: "First attempts at setting up the recorder as per instructions were fruitless, since schematic diagrams supplied were incorrect as to part numbers, tube types used and wiring of components. The legend giving values of resistors and condensers was incorrect in nomenclature with respect to the above mentioned schematic diagrams and the values given did not correspond to the actual values of the components." The memo went on to explain how, after tracing the circuits manually, technicians found a number of shorted condensers and resistors.[40] After Davies discreetly broached the issues with them, Harry Wells and D. E. George of the Department of Terrestrial Magnetism, which had supplied the recorders, explained that the design of the equipment was unnecessarily complicated but pointed out that "identical" equipment sent to Christmas Island had worked fine. They emphasized that DTM considered it essential for ionospheric equipment to be operated for a test period at the labs by the people who would use it at the field station—a procedure not followed in the case of Torbay, but strictly observed for all the other nine stations equipped and staffed by DTM in the preceding two years.[41]

Across the ionospheric stations controlled by the United States, Britain, and Canada, scarce parts and local modifications had created disjuncture after disjuncture between the stripped-down mimeograph depictions of machines in circuit diagrams and operating manuals and the thick material devices actually occupying the stations' Operating Rooms. According to Davies' description of the British ionosonde at Churchill, the equipment had been produced "at a time when availability of materials was a problem and under conditions in which almost any other electrical equipment had higher priority." "The results," he continued, "have not been satisfactory. Numerous criticisms of model #249 have come from US and Canadian as well as British operators."[42] Elaborating on the problems of reliability with the British equipment at Churchill, one of Davies' colleagues would later explain that "a number of circuit changes have been made by N.R.C. so that the present equipment is not identical with the original equipment. This equipment is the only one of its kind in this Dominion [of Canada], and perhaps in the U.S.A."[43] The equipment at Chelsea, built in 1940, suffered the same problems, having been "changed considerably from the original model so that no circuit diagram can be used for trouble-shooting, etc."[44] The DTM episode belied the nominal identity of ionospheric equipment. Operators needed to cultivate not only judgment, but judgment in relation to specific, idiosyncratic machines, eliding the distinction between human and technical error.

Unable to control personnel, Davies turned to the prospect of new machines as a solution. His program for investigating "obscure and exceptional ionospheric conditions" had always been based in part on improved ionosondes. The frequency range of most of the existing recorders was too small to capture maximum frequencies predicted for 1947.[45] For the manual ionosondes in use, the technical process of creating h'f curves could not capture rapidly changing features of the ionosphere above most of Canada— especially the phenomenon known as "sporadic ionization." That condition had to be accounted for in forecasting models for the polar regions and was "*defined* as traces seen on h'f ionospheric records which show rapid or erratic variation in frequency or vertical height as compared with the normal region traces."[46] Now facing the acute problem of personnel, Davies pushed even more energetically for the adoption of new panoramic ionosondes at all Canadian stations.

The panoramic ionosonde, an elaboration of the automatic ionosonde, was created in 1933 by T. R. Gilliland precisely to deal with the problems of unreliable manual records.[47] During World War II, researchers at the National Bureau of Standards began adapting it to deal with the shortcomings that even automatic equipment posed. Rapidly changing ionospheric conditions could confound automatic recorders as they drew photographic film across the beam of light. The new device added two "panoramic" displays, which automatically plotted the ionogram on the oscilloscope. A camera unit, triggered by operators watching a second display, recorded the images on the oscilloscope as a series of snapshots. Equipped in this way, an ionosonde was capable of operating at 15-minute intervals around the clock (as the sounding schedules of postwar operations demanded).[48] Davies had been acquainted with these recorders from his time at the Carnegie Observatory in Huancayo, Peru. In early 1947, he and Meek listed the virtues of the instruments, particularly for the Canadian program's new focus on tracing causes and effects: "These recorders present continuous photographic records of the ionosphere. Present recorders permit only intermittent measurements. Continuous measurements will give valuable knowledge of the cause and effect of movement and change in the ionosphere. . . . There are obvious advantages to having a uniform type of panoramic recorder at all Canadian stations."[49]

The new recorders would standardize critical parameters across American and Canadian stations, something that Davies considered important for the international status of RPL's program. But in the area of people and machines, they would have two important effects. By automating the drawing of graphs, RPL hoped, the devices would eliminate the need for careful judgment. By standardizing the devices, RPL could train personnel destined for isolated stations using identical equipment located at more accessible points, and requiring a more limited set of skills.[50] All that remained was to determine what "standardization" meant.

Standardization

On one view, standardization is about universality. It helps make the world fit for science by making things the same, thereby ensuring that results obtained at one site can be compared with others produced a world away.[51] The question of who defines standards and who merely adopts them points to their place at the heart of hegemonies, even "consensual" ones. This perspective on standardization as a thoroughgoing homogenizing enterprise has been instrumental in both problematizing and answering the question of how knowledge and results travel and highlighting the often enormous amounts of work that go into producing the same results at different locations.[52] If taken too far, however, that focus on replication threatens to obscure another facet of standardization: how it creates similarities while simultaneously creating the conditions for legitimate difference and therefore creating another kind of identity. In this way, standardization is only partly about identity in the sense of making things identical. Instead, it creates "conditions of instrumentality"—the constraints on the material culture of instruments that grant them validity *qua* instruments in the investigation of given effects.[53]

Throughout the latter half of the war, while the efforts for worldwide coordination of meteorological data were suspended, ionospheric researchers meeting in London and Washington had argued for standardizing the machines and practices of ionospheric physics.[54] Their work operated under the banner of "coordination" and pushed into the most arcane details of wartime ionospheric physics and the handling of records. Item by item, delegates had gone through the measurements and techniques that formed the most basic work of the ionospheric laboratories—how to measure ionospheric layer heights and critical frequencies, how to draw the world contour maps that limned the morphology of the ionosphere, how to transform vertical incidence data from ionospheric recorders into oblique incidence data for use in wartime communications, and how to produce the ionospheric predictions that formed the basis for Allied communications throughout the theaters of war. Day after day, delegates had hammered out compromises and consensus wherever they could, "disposing" of methodological discrepancies in a process that would continue through meetings over the next year.[55] But apart from a passing reference to a preferred *type* of machine, no direct reference to the devices of ionospheric research appeared anywhere in the official records. By early 1944, the delegates had rewritten many of the most widespread techniques of their discipline and had left machines virtually untouched.

Toward the end of the war, the issue of machines emerged explicitly after Newbern Smith at the US National Bureau of Standards recommended that "uniform" results be produced at stations across Canada and the United States. In correspondence, Smith had suggested that the best practices developed by the ionospheric groups in England, Australia, the United States, and Canada should form the basis of a prototype ionosonde to be developed cooperatively by the United States and Canada and turned over to a commercial company for production.[56] In late October of 1945, the principals of the Canadian and American programs assembled in Ottawa to see what could be done.

H. R. Smyth, whose ionospheric research stretched back to the 1930s and who was now engaged in developing the prototype Canadian panoramic recorder, summed up the domestic Canadian situation:

> For quite some time, there have been discussions in Canada about the necessity for standardizing ionospheric equipment throughout the world. Some individuals are in favor, others are not.

Wary of problems with logistics and manufacturing involved in such an enormous effort, Smyth suggested priorities:

> If we attempt to standardize on actual equipments as far as components are concerned, it might be advisable to try to standardize on sweep, pulse length to be used, speed of sweep, etc. In that way, records received from the different stations would be comparable. Items, however, which can be standardized without difficulty are: (1) Type of record for an ideal equipment. (2) Frequency sweep desired, logarithmic or linear. (3) What sort of marker should be placed on the sweep.[57]

Smyth's proposal was significant for its emphasis on images over material details. Putting aside the question of a homogeneous material culture for global ionospheric research, it envisioned instead a straightforward *formal* standardization of the products of the ionosonde.

Ever since the discovery of the Longitude Effect had driven the expansion of the ionospheric stations, Frank Davies and his group had struggled with the problem of ensuring that the different stations (now controlled by five different organizations) produced uniform graphs. Inexperienced personnel and the complexity of Canadian records combined to create intense anxiety about the records' integrity.[58] How to ensure that Air Force radar personnel and Department of Transport telegraphists would interpret the traces of the CRT in the same way? How to make sure that equivalent interpretations were rendered faithfully and identically on the logarithmic paper of the h'f curves? Smyth also expressed the concerns of Davies and his group. When approaching standardization, the Canadian concern was about the *form* of the records—logarithmic versus linear scales, panoramic versus traced, the appearance of frequency and height overlays, and so on. Davies and his colleagues focused on having the machines achieve what they were trying to achieve through training and personnel: the reliable and uniform production of records.

Not everyone agreed. Harry Wells, head of the Carnegie Institution's Ionosphere Section during the war, saw the emphasis on machines and images as misplaced. Working with Lloyd Berkner, Wells had built up the first network of automatic ionospheric observatories. During the war he had presided over another standardization debate: the one that raged after military officials had tried to coordinate all Allied ionospheric research centrally. The conclusion officials reached in that debate was that the pen-and-paper techniques of the interpreters would be matched to different classes of instruments in order to produce identical results. This was precisely what had allowed machines to remain unimportant in the initial discussions about standardization. Wells, pointing to the expense and effort to get identical machines, now extended that emphasis on uniform interpretation to the cathode-ray tubes of the field stations[59]:

> It seems to be, in the present stage of development, practically an impossible task, but certainly an uneconomical task to try to standardize on equipment. We can certainly achieve standardized *interpretation* of what the equipment turns out. So that, if we have three or four different types of recorders operating side by side and operated by entirely different groups of individuals with different training, they are all going to get the same results when converted into terms of information which are required for military or civilian application, and the final answers are going to be the same.[60]

Newbern Smith, the author of many of the fundamental ionographic techniques of the 1940s and the 1950s, was less concerned with either displays or interpretation. Smith had been the creator of the transparency method, in which "transmission curves" were printed on transparencies with frequency scales identical to those of ionograms and researchers could then lay the transparency directly over the photographic records. By reading off the points of tangency and intersection with the trace, radio engineers could determine high and low path frequencies and maximum usable frequencies directly from the graph's abscissa. In combination with models and statistical records from ionospheric physics, the techniques embodied in the transmission curves would be used to generate frequency prediction charts issued to radio operators across the Western Hemisphere in the late 1940s.

Smith had the lessons of the transparency method in mind as he approached standardization in October 1945. Interpretation for the purposes of data reduction did not interest him much. Operational needs, he believed, might be "lifted" out of the scientific work of the propagation laboratories, but they should not drive the priorities of ionospheric research, which involved the identification of phenomena. As long as the basic phenomena appeared on the records in some form, adjunct modular technologies could always transform them into useful and standardized information. "It does not matter too much," Smith argued, pressing his case for a kind of standardization of effects, "whether records are displayed as logarithmic or linear sweeps. It does matter, however, . . . what the ratio of the received signal to the atmospheric noise might be and what are the directional patterns of antennas." Smith therefore urged his colleagues to consider "standardization of receiving power, receiver sensitivity, and antenna directivity, and consider these as the more important things rather than concern ourselves over certain types of display."[61] Smith's ultimate goal was a kind of phenomenal standardization in which the same phenomena would be sure to appear on the records, whatever their specific form.

What lay behind these different views of standardization? Plagued by the problem of reliable personnel in the North, Davies and his group were concerned about the integrity of the ionogram as a visual record. Wells, having created the first network of automatic sounders, faced constant variations in those improvised machines, and sought to compensate for them through the malleability of people and practices. Smith, heading the largest center for the collection of ionograms from around the world, saw differences in interpretation and analysis every day and was far more concerned that all the records were capturing the same phenomena. The debates about instrumental standardization in North American ionospheric physics show us the manifold meanings the term held for researchers in the late 1940s—a comprehensive material identity between instruments designed to correct the variation and improvisation of wartime tools; an interpretive-instrumental matrix that compensated for material variation through human practice; or a phenomenal standardization that reduced ionosondes to black boxes judged by their ability to capture effects. The arguments were more about constraints than about hegemony or crushing uniformity. Difference proliferated in the interstices between these versions, as well as in the legitimized spaces that standardization created.

Ultimately, Wells' proposal came closest to the final result. The panoramic recorder eventually designed by the National Research Council and developed by the Canadian Marconi Company would be "very similar in basic design to the US set."[62] But neither it nor its visual products would be identical to the American device, allowing the frequency range of the Canadian sounders to be tuned to the high-latitude phenomena that would set them apart.

Throughout their deliberations about uniform equipment, Davies and Meek had concluded that the proposed outward standardization of machines and people—even if carried out rigorously across the Canadian stations—would go only so far. To ensure that the considerable expense of training and equipment wouldn't be wasted, they made it clear that "too much emphasis cannot be placed on the maintenance and *calibration* of equipment and its correct operation."[63]

In the same document in which Davies and his RPL colleague J. C. Scott spelled out the necessity of the panoramic recorder as a solution to many of RPL's problems, they also laid out the metrological context into which these machines and their operators should be introduced: "The following calibration and test equipment is required at each ionosphere station."[64] They then went on to list a series of signal generators, output meters, and analyzers that would help ensure the validity of the station data. To ensure its validity *across* stations, allowing researchers to map machinic effects spatially and tie them to national territory, Davies and Scott further proposed a "traveling calibration party."[65] Equipped with standard field-strength meters, vacuum-tube voltmeters, and signal generators, the calibration party "should include a qualified radio engineer with technical assistants responsible for the periodic inspection and calibration of equipment on all ionosphere stations. This engineer should be authorized to instruct station personnel in the proper upkeep of equipment and in its correct operation. He should also supervise the installation of equipment."[66]

The role of leader of the calibration party fell to Jack Meek, Davies' second-in-command at OIC/6 and a central figure in the laboratory's future research. Meek, who had served in the Royal Canadian Navy during the war, had been attached to OIC/6 because of his background and interests in radio physics. In 1947, as the Radio Propagation Laboratory was taking shape, he had led the research team of the Mobile Observatory, a modified railway car, in making ionospheric measurements across the auroral zone. In the spring of 1949, he traveled north again to Baker Lake, a tiny outpost about 800 kilometers inland from the northwest shore of Hudson's Bay, originally established as a fur trading post in the early twentieth century and often cited as the geographic center of Canada. There a Department of Transport station had recently taken delivery of a panoramic ionosonde on loan from the US National Bureau of Standards. The officer in charge, E. E. Stevens, would later join RPL's sibling, the DRTE Communications Laboratory, as the head of its Arctic Research division. In the spring of 1949, however, a month before Meek's visit, Stevens reported his progress with the instrument to Ottawa as follows: "Except for getting the clock and print timer synchronized no difficulty was experienced. Since that time a few sweeps have been missed due to our inexperience with the camera operation, but we now have a system in which I expect little further trouble. A small piece of film showing some of our first sweeps is enclosed."[67] (See figure 2.5.) The film showed the first panoramic ionograms ever produced

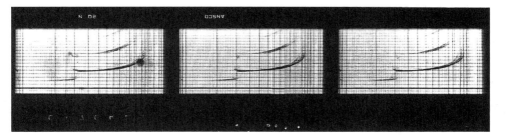

Figure 2.5
The first ionograms produced in February 1949 at Baker Lake, Northwest Territories. Produced by E. E. Stevens, these records were the result of an arduous calibration program designed to ensure their value. Source: Letter from E. E. Stevens to W. B. Smith, February 15, 1949, Library and Archives Canada, Record Group 12, volume 1641, file 6802–133. © Government of Canada. Reproduced with the permission of Library and Archives Canada (2016).

in the Canadian North. Ten days later, Stevens sent an urgent telegram to Ottawa: "please rush three only 801a tubes STOP one of the two spares defective and the last one in operation STOP these are for automatic recorder."[68]

When Meek arrived, two weeks later, the equipment was still functioning, but the 16-millimeter camera mechanism was operating continuously rather than once per sweep. As a crude solution to the problem, Meek and Stevens proposed placing adhesive tape on the cam responsible for the camera circuit. As they tried to run the ionosonde for several days (as scheduled, and well within its rated limits), their modification caused the camera's motor to become worn and the photographic equipment to misfire. Replacing the equipment with the spare produced identical results.[69] During the short period of successful operation, Meek's visit helped coordinate the sometimes delicate settings responsible for high-quality ionograms—pulse rates, pulse lengths, scales, and sweep lengths. Stevens would later report that the resulting records had been "much more comprehensive than those obtained with manual equipment, especially where the layers are changing rapidly as is often the case at Baker Lake."[70] Ionograms taken the following month bore witness to his views.

Beneath these images, however, lay the recalcitrance of the instrument. With unreliable electronics and inadequate specifications, calibration could go only so far. The validity of much of the ionospheric data would have to be taken on faith, backed up by the promises of metrology. The full discipline that Davies and his RPL colleagues had envisioned for the ionospheric stations would have to wait for the wide-scale reliability programs that would engulf large elements of the Canadian defense establishment in the years to come.

Conclusion

For Frank Davies and his group, nothing less than the definitive links between radio failures and Canadian geophysics rested on the reliable production of ionograms. They had turned to the panoramic ionogram as a solution to the problems of control at the ionospheric stations. The machine promised to capture the features of the turbulent ionosphere above Canada's North at the same time it solved the deep problems of men and machines that plagued the ionospheric project in those years. Producing reliable records with unreliable machines demanded expert judgment, but the conditions of both machines and people in the postwar North called their reliability into question. It was the idea of these machines in the field—cut off from proper supervision, isolated from adequate supplies, handled by inexperienced personnel who themselves were unreliable when placed in the Northern landscape—that preoccupied Davies. For that reason, one of the first acts of the Radio Propagation Laboratory was to request new instruments and a new type of visual record. Controlled by five different organizations, plagued by a constant turnover of personnel, and situated in harsh environments, the field stations represented a machinic order as problematic as the short-wave communications networks they aimed to improve.

Far from seeing standardization as an exercise in subordination, the newly created RPL welcomed it. Standardized instruments would make their results count within the wider synoptic projects of the United States and Britain. The terms of standardization—what was being standardized and how—mattered crucially. Were machines around the world to be identical to one another, down to the last resistor? Was an identity of output enough? Or was it sufficient for different machines to capture the same phenomena? These material, formal, and phenomenal understandings pointed to the history of standardization in the service of autonomy rather than hegemony, a polysemous standardization underwriting the project of identity. Materials, interpretations, and phenomena were all candidates for sameness. Difference proliferated in the interstices between these versions, as well as in the legitimized spaces that standardization created. For Davies and his group, standardization promised to create the very conditions under which significant difference would emerge. But by reliably capturing the complex and unusual phenomena above Canada, the new machines pushed anxieties from production to interpretation. Where attention and anxiety had originally been focused on the machines of the Operating Room, they now shifted to the desk of the Office Proper and to how the images should be read. That shift of a few centimeters on the station blueprints would launch an even more ambitious program to legitimize the distinctive natural order above Canada and its links to the nation's communications failures. In the process, it would radically redefine the reading technologies of ionospheric physics.

3 Reading Technologies

In the early spring of 1956, one of the more remarkable maps of the Cold War surfaced in a report probably never seen in Washington or Moscow. It was produced by the government of the Yukon Territory, the smallest and westernmost of Canada's two Northern territories.[1] Bordering Alaska, the Territory (a remnant of the Hudson's Bay fur-trading empire) had become part of the Northwest Staging Route during World War II and a major geopolitical interest immediately afterward. The map in question exploited military cartographic conventions, where the breadth of arrows represented the size of invading forces and appeared to show plans for defending the Yukon against invasion by the Soviet Union. (See figure 3.1.) As Soviet forces swept in from the Russian Far East, American forces countered from Alaska and the Pacific Northwest. Two Canadian contingents (the tiny arrows at the bottom right) provided what support they could. What was remarkable about the map, then, is that it did not represent military forces at all, but rather the reception quality of short-wave radio broadcasts into the region. Its author, the Yukon's commissioner, F. H. Collins, worried that the North was becoming "an interesting battleground of Soviet and American ideologies through the medium of radio, while Canadian viewpoints are completely absent."[2] The brief that accompanied the map turned radio failure into a polyvalent metaphor for the cultural and geopolitical threats to the Canadian nation—an electromagnetic instantiation of hostility and vulnerability. In the years before and after, those views were echoed again and again as observers claimed that the situation of Northern radio threatened Canadian national life and embodied Canada's struggle to survive as a national and cultural entity.[3] Situated among those narratives, the map formed a visual argument for just how extensively the project of Northern radio was shot through with the politics of the Cold War.

The previous two chapters focused on the machinic orders that underlay images such as this—the geographic distribution of radio blackouts and

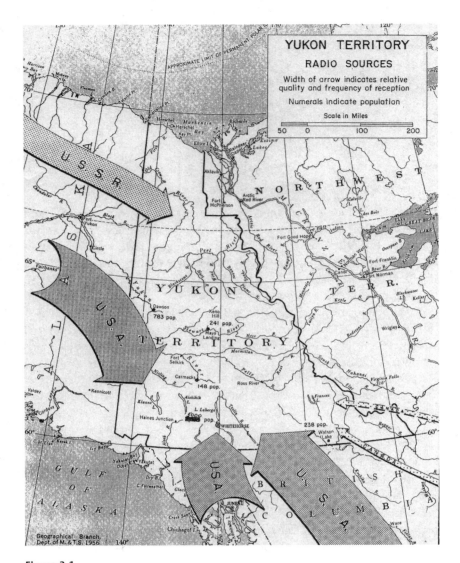

Figure 3.1

A map of radio transmissions into the Yukon. The arrows indicate the source of the transmissions entering the Territory; their breadth indicates reception quality. Note the two small arrows in the lower right representing Canadian broadcasts. Source: F. H. Collins, "Radio in the Yukon Territory: A brief presented to the Royal Commission on Broadcasting," A Brief Presented to the Royal Commission on Broadcasting, April 1956, Library and Archives Canada, Record Group 41, volume 127, file 5, part 2. © Government of Canada. Reproduced with the permission of Library and Archives Canada (2016).

instrument behaviors, and their significance in the early years after World War II. Confronting anxieties around wartime radio and detection failures in the North Atlantic, the Radio Propagation Laboratory had translated those breakdowns into elements of a late-war political program rooted in visions of a distinctive natural order and aimed at reclaiming the Canadian North physically and epistemically, using it to elevate the postwar nation and to reshape its status and identity. After the war, those anxieties had quickly pushed beyond radio sets and direction-finding equipment to the machines of ionospheric research itself. Handled by unreliable operators no longer under the supervision of Frank Davies and his group, the instruments could not be counted on to capture the distinctive geophysical effects that their project was built on. Humans, instruments, and communications were all unreliable in the postwar North. And to secure them, RPL had turned to new standardized machines that would capture the full complexity and dynamism of the high-latitude ionosphere and would underpin the broader attempts to create a distinctive scientific identity for the nation.

This chapter explores how those concerns were pushed further still, beyond physical devices and into another set of "technologies." Radio communications and ionospheric instruments were not the only things that broke down under the effects of hostile Northern nature; the reading technologies of ionospheric physics—the very ways in which the scientific images were seen, analyzed, and apprehended—broke down as well. Ultimately, the failures that surrounded Northern communications, and the anxieties and arguments that accompanied maps like the one of the Yukon, motivated a program to change the practice of seeing in ionospheric research, the act that linked Northern nature to the machinic order of high-latitude communications and that stood at the heart of RPL's project.

There are precedents for investigating the relationship between laboratories and their visual products.[4] Drawing on recent literature, for example, visual representations might be placed on par with concepts, theories, social relations, and material processes as the products and tools of a more fundamental activity: isolation, interrogation, and intervention into natural processes.[5] Or else images might ground the material, practical, and epistemic cultures of the laboratory, serving as critical resources in arguments about the reliability of experimental practice, the validity of instrumental arrangements, and the soundness of theoretical conclusions.[6] Further still, the laboratory itself might be understood as a site rooted in the transformation into paper of phenomena made manifest within its walls.[7] But whether we see the laboratory as interrogation chamber, as inscription factory, or as something in between, we recognize that its relation to images and the power they seem to hold derives principally from one thing: the claim of the laboratory to make some physical aspect of the natural world—whether through modeling, manipulation, or mimesis—enter its controlled spaces.

The ionospheric laboratory of the early 1950s could make no such claims. Ionospheric physics in the period after World War II belonged to the sciences of the eye.[8] Like those sciences, it depended on specific technologies of sight and reading to carry out its work; unlike many of them, however, it had no direct control over its object of study.[9] Whereas the botanical specimens or tissue samples of plant biology or histology could be manipulated directly, ionospheric phenomena remained remote. Instead, images of the ionosphere—ionograms—furnished the "working objects" of the discipline.[10] Because of this, there were at least two ways in which ionospheric investigations could break down. First, specific phenomena could fall outside the standardized working objects of the discipline, falling into the cracks between accepted phenomena and authoritative effects. Second, the accepted techniques for reading the images, for putting them to theoretical and practical work, could founder on anomalous records. Both of these things happened in the late 1940s in high-latitude research. Initially, turbulent ionospheric conditions, untrustworthy workers, and malfunctioning, idiosyncratic machines had generated problems in producing images that faithfully captured the distinctive phenomena that Davies' group, now renamed the Defence Research Telecommunications Establishment, was after. To overcome those difficulties, Davies and his staff had turned to a new standardized machine, the panoramic recorder. But the new records created a different problem. By capturing faithfully the complexity and turbulence of ionospheric conditions, they produced ionograms whose phenomena fell outside the working objects of the discipline, making those records effectively unreadable.

In this chapter, I examine the efforts to make the high-latitude ionogram legible, tracing the effects of that new legibility into wider, resonant views of the relationship between the North and communication failures. The chapter opens by exploring the transformations in the way the high-latitude ionogram was read. As part of its project to improve Northern communications and articulate a Northern identity for the country, DRTE's Radio Physics Laboratory mounted a novel approach to investigating radio disruptions. Rather than follow the statistical correlations that dominated the discipline, they set out to trace the intricate chains of cause and effect that linked Canada's peculiar geophysics of magnetic storms and auroral displays to radio blackouts. The problem Davies' group faced was that the same geophysical phenomena that disrupted radio communications, the very phenomena at the core of their research, also obscured the most distinctive features of high-latitude ionograms, making the records unreadable using standard techniques. Led by one of its founding members, Jack Meek, the laboratory began working out a set of reading regimes that would make even the most "disturbed" high-latitude ionograms readable for the first time. Those techniques would be used by DRTE researchers to connect high-latitude atmospheric phenomena to other Northern geophysical and radio phenomena in ways that had previously been impossible. The chapter then turns to investigate how those alternative reading techniques created associations between the North and radio communications that resonated far beyond the laboratory. By linking Northern geophysics and communications disruptions through this new way of seeing ionograms, the laboratory furnished visual arguments for how the very things that defined Canadian northernness, and with it Canadian identity, threatened reliable communications. In doing so, the laboratory's work fed back into broader cultural narratives that had supported its research in the first place. Turning to arguments beyond the ionosphere, the final portion of the chapter traces out a number of sites where RPL's linking of Northern nature and technological failures were rehearsed again and again. In order to understand these developments, though, we must come to grips with the general understandings of the ionogram within ionospheric research, the disciplinary traditions they drew upon, and how they made the image speak to the interests of the early Cold War.

Extraordinary Traces

Six years before the Yukon radio map appeared, Davies' group had produced its own illustration of radio transmissions in the North. The map showed how radio disruptions affected Canada disproportionately, spreading further south into the country than into any other nation in the Western Hemisphere. (See figure 3.2.) For Davies' group, a first-order geophysical explanation accounted for the effects. The northern auroral zone sliced through the path of short-wave transmissions from southern Canada, disrupting them to the point of sometimes generating complete blackouts. Since the middle of World War II, researchers had tried to overcome those disruptions through more and better data from high latitudes. Davies' solution, however, broke with those suggestions. Shortly after the war, his group moved to abandon statistical correlations and instead focus on models that emphasized physical causation. The approach of the lab quickly moved from the aggregate to the particular, and from the normal to the disturbed ionosphere. RPL was interested in the complex geophysical phenomena

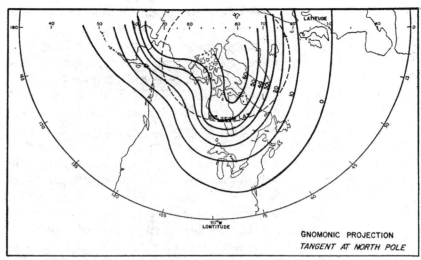

FIG. 7—AVERAGE NUMBER OF HOURS OF BLACKOUT PER DIS-
TURBANCE DURING NINE ABSORPTION DISTURBANCES OF
1949–1950

Figure 3.2
Meek's mapping of radio disruptions in Canada, showing how the country experi-
enced more, and more severe disruptions, than the United States or Britain. Source:
J. H. Meek, "Ionospheric Disturbances in Canada," *Journal of Geophysical Research*
57, no. 2 (1952): 177–190. Reproduced with permission from American Geophysical
Union.

that produced sudden ionospheric disturbances and the radio disruptions
associated with them. By tracing the geophysical histories of specific ionospheric disturbances, they hoped to identify their origins and to predict
when and where they would happen and how they might unfold.

In order to capture those ionospheric conditions graphically, they had
turned to ionograms. The power of the ionogram lay in two reading regimes
that had grown up around the image in the 1920s and the 1930s. Through
a series of mathematical and graphical developments before the war, ionospheric physicists had learned to interpret the graphs backward, as it were,
through magneto-ionic theory, thermodynamics, and cosmic-ray physics, to
explain ionospheric structure, formation, and dynamics. Using those techniques, they hoped to develop models of the ionosphere that both detailed
its vertical structure and predicted its future behavior, testing these by returning once again to the contours of the graph. Radio researchers, on the other

hand, read the images forward, through the laws of optics and through a series of propagation theorems that converted quantities measured directly on the graph into maximum usable frequencies, optimal working frequencies, and skip distances for communications circuits.

Both of these ambitious readings of the ionogram were built on a more basic act, however: what contemporary manuals simply called "interpretation." Carried out on the largest scale by field-station technicians around the world, interpretation was designed to momentarily collapse the possible meanings of the image into a single standard and authoritative reading that could be circulated globally.[11] It did this in two phases. First, observers parsed the features of the ionogram according to a standard taxonomy, breaking down the image into the various ionospheric layers—E, F1, and F2—and then into the phenomena that might be associated with each— Sporadic E, M, and N-type echoes, oblique reflections, and so on. This served the interests of morphology, but the ultimate purpose, the one toward which the entire interpretive act was directed, was to identify a single graphic feature known as the "ordinary trace."[12] (See figure 3.3.) That trace was the object of the second phase of interpretation, known as scaling. Although it was generally included as part of the interpretation process, scaling involved the measurement of two characteristic features of the ordinary trace—its critical frequency and its minimum virtual height—that furnished the raw data for ionospheric models and frequency predictions. Although the same measurements could, in principle, be recorded for the other major trace on the ionogram (known as the extraordinary trace), the characteristics of the ordinary trace dominated the reading of ionograms and the analyses based on them.

To teach people how to scale ionograms, manuals began by presenting an "ideal type" stripped of everything but the ordinary trace and suitably schematized for pedagogical purposes. (See figure 3.4.) Many scientific disciplines have used archetypes—representations of phenomena that are not, or cannot be, realized in any individual example, but that contain essential elements from which all possible examples might, in principle, be derived. The ideal type or *Typus* was a subset of the archetype.[13] It was defined by its idealized status in relation to some criterion—beauty, for instance. When the eighteenth-century Dutch anatomist Bernhard Siegfried Albinus, for example, set out to represent the human skeleton, he chose both materials and modes of representation that he believed brought out its beauty and perfection. Albinus' ideal type was a medium-sized, well-proportioned male skeleton that he had painstakingly cleaned, reassembled, posed, and then

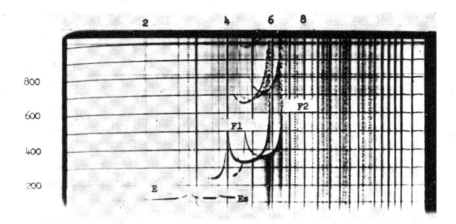

Figure 3.3

A typical mid-latitude ionogram. The mirrored curves represent the ordinary trace (left) and the extraordinary trace (right). The first of these was the focus of "interpretation" practices. Source: "IGY Instruction Manual, the Ionosphere: Part I, Ionospheric Vertical Soundings," ed. W. J. G. Beynon and G. M. Brown, *Annals of the IGY* III (1958). Reprinted with permission of Elsevier.

"traced" from a distance, using a double-grid system that imposed a rigid perspective. Features of the skeleton that were less than perfect were "mended" while remaining, in his view, true to nature.[14]

What pushed the ionograms used in instruction manuals into the category of ideal type, rather than mere archetype, was their legibility: they were perfectly readable. The minimum virtual height of the trace was read off the ordinate axis; the critical frequency was defined as the frequency to which the main portions of the trace became asymptotic. Idealized ionograms contained nothing to interfere with those two measurements or to call them into question. Scaling rules were then based on comparisons of actual ionograms with the idealized models, which singled out worthy and useful phenomena in ionospheric physics. Instruction manuals went on to emphasize the importance of having identical scaling procedures at all stations.[15] True, multiple interpretations were possible within the bounds of atmospheric physics and radio engineering proper; true also, the uniform act of interpretation itself might be shot through with theoretical presuppositions from both fields. But the act that manuals of the postwar period deemed "interpretation," it was argued, had to furnish an analytical middle ground from which physics and engineering could each move

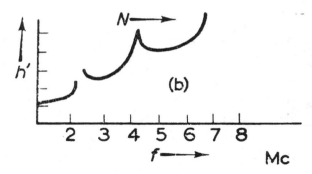

Figure 3.4
An idealized ionogram. The image, used to train observers during the IGY, has been stripped of everything but the all-important ordinary trace. Source: "IGY Instruction Manual, the Ionosphere: Part I, Ionospheric Vertical Soundings," ed. W. J. G. Beynon and G. M. Brown, *Annals of the IGY* III (1958). Reprinted with permission of Elsevier.

outward, a terrain as free as possible of idiosyncratic readings and personal variations.

Ionographic analysis therefore turned on this identification of the ordinary trace and the precise measurement of its characteristic features. Working in remote field station offices, staff were expected to recognize this trace in the relative confusion of an actual ionogram. And it was there, in the offices of the Canadian field stations and in the act of interpretation, that the standard techniques quickly faltered.

Reading at High Latitudes

The standard practices for reading the ionogram had been based on the relatively simple ionograms produced in temperate latitudes, where British and American researchers had first developed interpretation techniques.[16] But what, some researchers asked, was an interpretive system based on distinct traces, ordinary waves, and discrete frequency and height measurements supposed to make of something like the ionograms produced at Baker Lake, in the Canadian Northwest Territories? (See figure 3.5.) These "anomalous" records, obtained during ionospheric storms or under the frequently turbulent conditions at high latitudes, were aberrations in the standard scheme. Since they had no well-defined ordinary trace, they were unreadable under the standard practices. But they characterized the ionograms produced

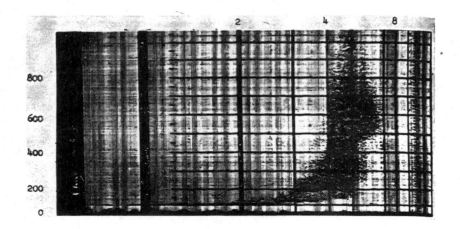

Figure 3.5

A severe case of "spread echo" (Baker Lake, 1400, August 4, 1955). Such phenomena made high-latitude ionograms unreadable under the standard rules of interpretation, even though they characterized records produced at many Canadian ionospheric stations. Source: W. J. G. Beynon and J. W. Wright, *IGY Instruction Manual: The Ionosphere* (Pergamon, 1957). Reprinted with permission of Elsevier.

in many parts of Canada, the very records that were central to RPL's project.

E. E. Stevens, who produced the first ionograms taken at Baker Lake in 1949, would later explain that the high-latitude records at Canadian stations posed two types of problems. First, their structure was more complicated than mid-latitude ionograms. The strength and direction of the magnetic field in the Canadian North produced additional traces on the graphs (most notably a third, characteristic trace called the Z-trace), which made the identification of the ordinary trace more confusing. "At certain times," Stevens noted, "high latitude ionograms may resemble those obtained at the lower latitudes. More frequently however, they exhibit a complicated structure that is characteristic of the high latitude ionosphere." Throughout the 1950s, RPL researchers would assiduously compile and analyze these pictures, eventually gathering them into codices of high-latitude effects that defined, as Stevens called it, the "Canadian ionosphere."[17]

The second problem was that the occasional phenomena on which standard interpretations foundered seemed to *characterize* images at high latitudes. Spread echoes like those shown here in figure 3.5, intermittent at temperate field stations, were normal and frequent at many high-latitude stations. They were joined by additional features—night E layers, brush traces, and polar spurs—that frustrated attempts at standard interpretation. Because of their relative infrequency in middle latitudes, these troublesome phenomena were labeled "sporadic." By obscuring the cherished ordinary trace, the central working object of ionographic interpretation, they impaired the readability of high-latitude records. Stevens: "While it is clear that comparatively few ionospheric phenomena are unique to high latitudes, unusual ionograms are encountered more frequently and anomalies such as spread echo, blackouts, and E_s [sporadic E ionization] can cause far more uncertainty in scaling than at moderate latitudes. This, then, is the principal difference found with high-latitude soundings: *more doubt exists in the recorded data, making the ionograms more difficult to scale.*"[18] Echoing Stevens' point more sharply, one US official claimed that Canadian researchers faced "probably the most difficult problem in the world, from an ionosphericist's point of view . . . that of standardizing interpretation of records because your recordings are so greatly disturbed by the auroral zone."[19]

Gaping omissions in the daily tabulation charts of stations throughout the high-latitude regions testified to the awkward status of high-latitude ionograms.[20] To deploy the standard techniques of ionographic analysis in the early 1950s was to produce, to the extent that ionospheric storms and technical difficulties permitted, an authoritative reading of the ionogram; but it also reinforced the hold of a taxonomy of ionospheric effects and ionographic appearances that pushed entire classes of records and entire groups of researchers to the margins.

One of RPL's persistent concerns since its formation involved preventing the ionogram from sliding into the terrain of mere artifact. (See chapter 2.) That anxiety had originally circled around the Operating Rooms of the field stations, where technicians produced ionograms at 15-minute intervals, in rotating shifts, 24 hours a day. As control over the staffing of the field stations passed to the military, Davies' group found it harder and harder to secure experienced and conscientious technicians. Malfunctioning machines sharpened those concerns. The known effects of isolation, shortened daylight, and long winters further eroded the trust that RPL felt it could place in the stations' personnel and the data they produced. To help solve those problems, RPL had turned to the panoramic ionosonde. But while the new machine eased the problems of production, it ultimately shifted the site of anxiety within the small spaces of the field stations. By capturing reliably and in detail the complex and unusual phenomena above Canada, the new machines pushed anxieties from production to interpretation. Where attention and anxiety had originally been focused on the machines of the Operating Room where the ionogram was produced, the new records shifted worries to the desk of the Office Proper, where the ionogram was interpreted, and to the people and practices that occupied it. That shift of a few centimeters on the station blueprints would help radically redefine the technology of reading in ionospheric physics.

The category of "sporadic phenomena" helped drive that redefinition. No physical criterion linked the different effects that went under the title "sporadic." Instead, they were defined by their *appearance* on the ionogram, where they figured either as rapidly changing or erratic phenomena, or else as phenomena that appeared infrequently at middle latitudes. Since those phenomena characterized high-latitude graphs, they made the majority of records produced in the Northern polar regions, especially in Canada, either marginal or worthless. And because they made ionograms unreadable, those same phenomena frustrated attempts to link quantitatively the peculiar geophysical phenomena that caused them to the radio disruptions they engendered.

Starting in the late 1940s, researchers at RPL began developing ways to make even the most anomalous high-latitude records legible. Their response can be broken down into two major efforts. The first, taking place from the end of the war to early 1952, focused on a set of techniques for rescuing the ordinary trace piecewise from the confusion of sporadic phenomena. This quasi-archeological approach ultimately bowed to standard interpretation techniques, their immediate concern for the ordinary trace, and the measurement of its discrete values. In early 1952, however, as concerns about Soviet intentions in the Arctic intensified and the issue of Northern communications became even more urgent, researchers at RPL began developing a system that would incorporate high-latitude ionograms wholesale into the body of ionospheric data.

A central figure in both these attempts was Jack Meek, a sub-lieutenant in the Canadian navy during the war who had been assigned to OIC/6 because of his background and interests in radio physics. Along with Davies and J. C. W. Scott, Meek was one of the founding members of RPL. In 1947, as the laboratory began its operations, he had led a small research team to Manitoba and used a converted railway car to make mobile ionospheric measurements at positions across the auroral zone. His research focused primarily on the correlation of magnetic, auroral, and ionospheric variations. Those interests would lead him, in the early 1950s, to pursue a master's degree and a doctorate on the topic, both at the University of Saskatchewan. But already during the war, Meek had begun questioning the standard interpretive practices for ionograms and particularly their emphasis on the ordinary trace. His doubts came out of dealings with a phenomenon known as "Sporadic E" (E_s)—small clouds of intense ionization that could sometimes reflect radio waves near the ultra-high-frequency range, creating unusual propagation paths. Two features defined these sporadic layers. First, they appeared at irregular intervals both day and night, which meant that, unlike the normal E-layer, they were not caused by ultraviolet solar radiation. Second, they generally lacked a discernible critical frequency, which meant that they drew attention mainly for the way they interfered with normal measurements of the ionogram. Reacting against the practice of discarding measurements of sporadic E ionization unless the ordinary trace was present, Meek suggested using the *extraordinary* trace (whose separation from the ordinary trace was proportional to the strength of the magnetic field) to derive values for the missing trace. Similar attempts to recover obscured traces and to incorporate marginalized elements of the high-latitude ionogram into the accepted practices of ionospheric research would characterize his work in the two decades after the war.

Meek's postwar work followed his interest in characteristic types. In 1946, he drew up a new instruction manual for technicians that used a large number of actual photographs and detailed interpretations as examples. Out of the profusion of effects that appeared on high-latitude ionograms, Meek chose exemplars that would guide the minds of operators and allow for generalizations and comparisons. In a 1949 article in the *Journal of Geophysical Research*, he continued his attempt to move the bulk of high-latitude ionograms beyond the sphere of idle curiosity. His first step was to redefine "sporadic" phenomena. At temperate latitudes, "sporadic ionization" had been a catch-all term referring to phenomena that changed rapidly or erratically, as well as to phenomena that occurred infrequently at middle latitudes. But in polar regions, Meek observed, the presence of ionization whose intensity changed rapidly and erratically was *characteristic* rather than exceptional. "It is this [rapidly changing] phenomena," he announced, "which we shall call 'sporadic ionization.'" Having normalized the occurrence of the phenomena, Meek followed it with a visual taxonomy of sporadic effects. (See figure 3.6.) Using records taken at Churchill and Baker Lake, he divided his classifications between the highly structured traces of the F region and the more patchy and diffuse echoes of the E region. In the former he included all the effects—spread echoes, polar spurs, forked traces, oblique reflections—that he believed characterized ionograms in "northern regions."[21]

In the next two years, Meek elaborated the classification and coding systems into a new reading regime designed to deal with high-latitude

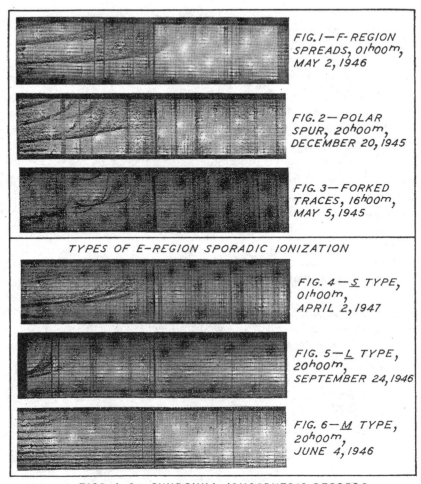

Figure 3.6

Jack Meek's visual taxonomy of sporadic ionization. The illustration shows the features that would come to define the high-latitude ionosphere—spread echoes, polar spurs, forked traces—along with letter codes for F-region conditions. Source: J. H. Meek, "Sporadic Ionization at High Latitudes," *Journal of Geophysical Research* 54 (1949). Reproduced with permission from American Geophysical Union.

ionograms. Together with his colleague C. A. McKerrow, he recorded it in RPL's new instruction manual for ionospheric observers published in 1951. The document began in the usual way: an introduction to the various regions of the ionosphere, examples of their appearance on ionograms, and a visual definition of the concepts of virtual height and critical frequency. In dealing with the question of interpretation, however, the manual strayed from traditional sources. It instructed observers to record critical frequencies for *all* wave traces: the ordinary wave as well as the magnetically dependent extraordinary and magneto-ionic components, which only appeared on high-latitude records. When the ordinary component was absent, it was to be *reconstructed* either through interpolation over a sequence of ionograms, or from higher-order reflections, or from the extraordinary component by way of frequency separation charts now included in the manual.

The real advance of the document for Meek and McKerrow, however, was the way it dealt with the effects of sporadic phenomena that had previously stigmatized high-frequency records. Building on Meek's earlier work, it divided effects according to ionospheric layer. But it also included for the first time a comprehensive list of *idealized* sporadic phenomena along with a description of when and how each might be scaled. In dealing with the spread echoes, "polar spurs," and forked traces associated with the F2 layer, for example, the manual referred its readers to an idealized series of records stripped to their essential components (figure 3.7), along with explanations of how critical frequencies might be recorded using the well-defined inside edge of certain parts of the trace. To this they attached a comprehensive system of letter codes (nineteen in all) that either explained why numerical values could not be produced, qualified the given value with the graphic circumstances surrounding its measurement, or explained the "reason for doubt" about its accuracy.[22] The monthly report sheets issued by the various ionospheric stations therefore contained a mix of scaled numerical values and qualifying symbols that allowed the data to wear their dependability on their sleeve, so to speak.

By 1952, the new system's effect on the ability to scale high-latitude ionograms was dramatic. A series of ionograms produced in January of that year at three locations—Barrow (Alaska), Kiruna (Sweden), and Baker Lake (Canada)—illustrates the change. Although all these stations were roughly at the same geomagnetic latitude (and therefore experienced similar ionospheric conditions), and although the American and Canadian stations used the same equipment (an NBS C-2 ionosonde), R. W. Knecht, a US specialist on the polar ionosphere, explained that "Baker Lake had no median determined by less than *16 values* whereas Barrow had a number of hours

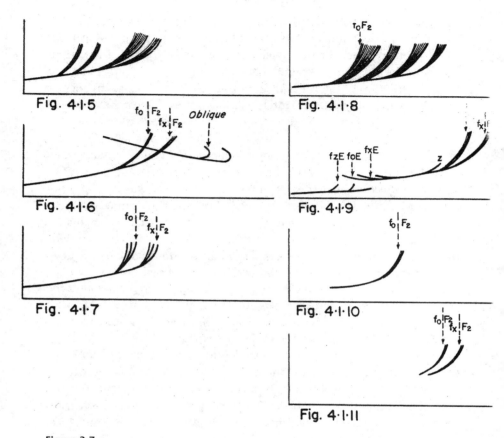

Figure 3.7

Idealized representations of sporadic phenomena. The illustration, produced in the 1951 RPL instruction manual, included polar spurs (4·1·5, 4·1·8) oblique echoes (4·1·6), z-traces (4·1·9), and forked records (4·1·7). Source: J. H. Meek and C. A. McKerrow, *Ionosphere Observer's Instruction Manual* (Ottawa: Defence Research Telecommunications Establishment, 1951). Reproduced with the permission of the Ministry of Industry, 2016.

with *no numerical values* and Kiruna falls somewhere between these two extremes." The discrepancy in the number of recorded values was attributable not to simpler Canadian records, but to the different interpretive and scaling practices used at each station. Citing four of the Canadian records obtained at Baker Lake (figure 3.8), Knecht explained that although the "original ionograms were undoubtedly clearer than the reproductions shown . . . CRPL [US Central Radio Propagation Laboratory] stations *would*

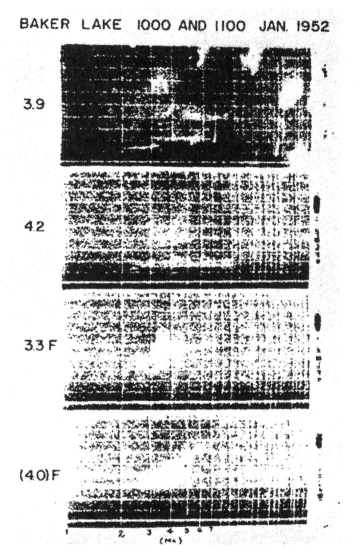

Figure 3.8
A series of ionograms from Baker Lake. None of these records would have been scaled by US observers, owing to the absence of the ordinary trace and the difficulty in reconstructing it. This image was published in R. W. Knecht, "Statistical Results and Their Shortcomings Concerning the Ionosphere within the Auroral Zone," *Journal of Atmospheric and Terrestrial Physics*, special supplement. Copyright: Elsevier, 1957.

probably not scale definite critical frequencies from the first 3 of these as did the Canadians." "This," he concluded, "is not a criticism of their practices . . . but rather an observation of still another problem: that of standardizing scaling practices among the various networks."[23]

In making so many high-latitude ionograms readable, Meek's techniques introduced a grave problem. The standard taxonomy and interpretive practices used to analyze temperate ionograms had at least made the job of the field workers simpler by excluding many complex records and avoiding the difficult judgments that accompanied them. With the new taxonomy and its descriptive symbols, uncertainties and potential misreadings multiplied. There was still the very real possibility that "many 'hidden' doubtful values" were making their way into the monthly tabulations undetected.[24]

Once again, Meek proposed a solution. During his work in 1952 and 1953, he had been interested in charting the temporal variations in three factors implicated in communication failures via the ionosphere: the various components of Earth's magnetic field, the position and intensity of the visual aurora above the northern horizon, and changes in the critical frequencies and heights of the ionospheric layers.[25] The aim was to correlate these factors, which reached their most potent expression above Northern Canada, but not to make that correlation the basis of a prediction model. Previous research had been based precisely on these correlations for forecasting disturbances (what Meek would later call "forecasting without understanding").[26] Those kinds of statistical models, Meek and his colleagues argued, tended to smooth out the great variations in ionospheric and geophysical conditions that typified high-latitude (read: Northern) phenomena.[27] Specific instances of ionospheric and magnetic storms therefore accorded poorly with the models. Instead, Meek envisioned a practice that laid out simultaneous but distinct geophysical phenomena alongside each other. (See figure 3.9.) The idea was to chart the precise parallel histories of individual ionospheric disturbances and associated geophysical phenomena. Researchers, it was hoped, could then suggest causal theories to predict the fine-grained machinic order that would result—specifically the location, extent, and magnitude of radio disturbances. That was a nearly impossible task with the regular methods for reading ionograms because the appearance of aurora and magnetic disturbances often coincided with a disappearance of the ordinary trace on ionograms, making it impossible to map quantitatively their simultaneous histories. To solve the problem, Meek had begun recording the entire range of frequency and height data from the ionogram, rather than just the discrete scaled values from the

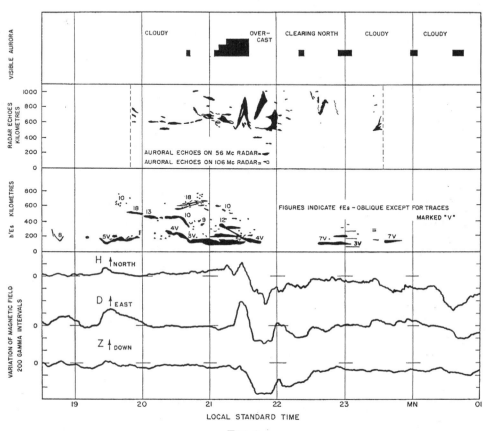

FIG. 1.

Figure 3.9

Meek and MacNamara's correlation of auroral, magnetic, and ionographic data. Correlations such as this would form visual arguments for the uniqueness of Canada's geophysical position. Source: J. H. Meek and A. G. MacNamara, "Magnetic Disturbances, Sporadic E, and Radio Echoes Associated with Aurora," *Canadian Journal of Physics* 32 (1954). © 2008 Canadian Science Publishing or its licensors. Reproduced with permission from National Research Council of Canada.

ordinary trace, laying them out alongside auroral and magnetic data to chart the history of a blackout.

Meek now moved to formalize that approach for high-latitude iono-grams. As the author of wartime and postwar instruction manuals for OIC/6 and then for the fledgling RPL, he had agonized over the interpretive quan-dary: whether to develop a system that would make interpretation simple and straightforward, and risk marginalizing many Canadian records, or to devise a system that would do justice to the complexity of high-latitude ionograms but would contribute to unreliability by requiring field workers to perform exceedingly difficult interpretations. Drawing on his own tech-niques for charting the history of disturbances, Meek drew up a graph dis-playing frequency as a function of time. Parsing the abscissa into 15-minute intervals (corresponding to the schedule of ionospheric soundings), he began recording information about critical frequencies over a 24-hour period. Distinct frequencies entered the graph as specific points, but (and here lay the force of the device) rather than requiring that all data be entered as discrete values or not at all, as previous techniques had done, Meek allowed the graph to record *ranges* of frequencies. Where an echo trace spread over the ionogram, field workers could simply record the entire range of frequencies it occupied. Furthermore, observers did not even have to identify which of the various traces (ordinary, extraordinary, or magneto-ionic) the frequency values belonged to, instead recording the information from the record in all its frustrating complexity. (See figure 3.10.)

The panoramic ionosonde had been a machinic solution to the prob-lems of judgment in producing ionograms. The *f*-plot, as Meek designated his device, was a graphic solution to anxieties about reading them. Whereas the act of interpretation had previously focused on the single ionogram, it now concentrated on a series of records illustrating the dynamic properties that set the high-latitude ionosphere apart. As one American observer com-mented: "Rather than analyzing isolated hourly soundings and trying to scale them according to low latitude rules, it was discovered that if the fre-quency data from each 15 min sounding were plotted *with a minimum of pre-judgment*, then meaningful hourly values could quite easily be obtained from the plot."[28] Researchers could now "see" the formation and disappear-ance of the sporadic layers, and the temporary blanketing of otherwise regular patterns in critical frequency data. As Meek's colleague Kenneth Davies explained, "The emphasis in interpretation [with the *f*-plot] is in estimating the value of a parameter in relation to the sequence of iono-grams before and after it, rather than in trying to see something equivalent to an undisturbed echo through the general confusion of a disturbance."[29]

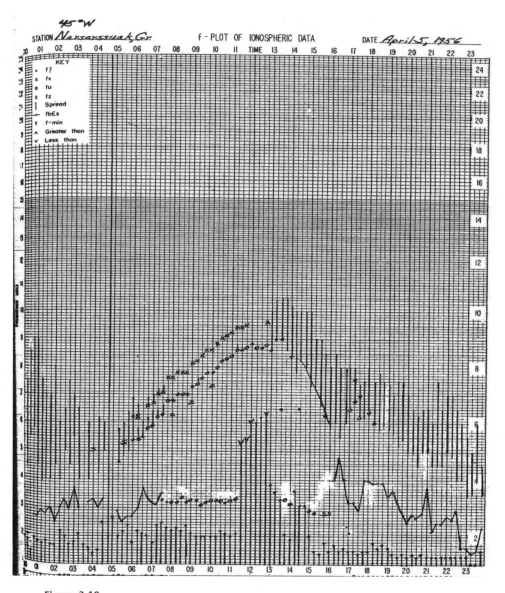

Figure 3.10
The *f*-plot. This graphic technique allowed ionospheric data to be recorded even in the absence of distinct traces. As this example using data from Greenland suggests, it would be adopted globally by the late 1950s, particularly for use on high-latitude records. Source: W. J. G. Beynon and J. W. Wright, *IGY Instruction Manual: The Ionosphere* (Pergamon, 1957). Reprinted with permission of Elsevier.

The reading regime that underlay the *f*-plot quickly became a standard tool in high-latitude research. The Special Committee on High Latitudes of the Union Radio-Scientifique Internationale adopted it in 1955 to combat the problems of interpretation at Northern stations.[30] Through correlations like Meek's, the plot tied the ionogram to a set of parallel visual products and graphic devices—auroral charts, radio echo graphs, and magnetographs. But an equally dramatic effect was the serial graphic products that the *f*-plot made possible.[31] Through the international adoption of these techniques, researchers began sketching out the characteristics and boundaries of the

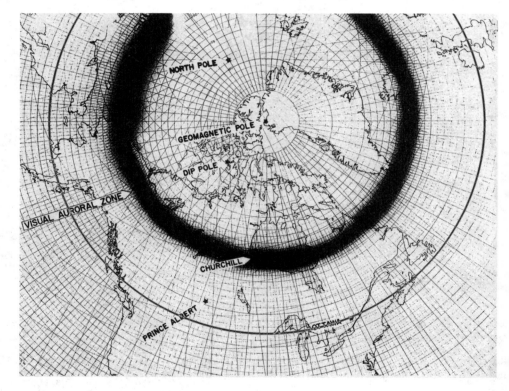

Figure 3.11

The high-latitude ionosphere. This illustration shows the geographic projection of the region (shown by the outer ring running through Ottawa). Because the southernmost station in Canada was located at Ottawa, this new construction meant that all data collected in Canada were considered "arctic" data. Source: *Alouette* (Ottawa: Communications Research Center, 1967). Reproduced with permission of the Ministry of Industry, 2016.

high-latitude ionosphere. The new scaling techniques and the f-plot created a high-latitude ionosphere to accompany the spatial distribution of radio disruptions and high-frequency blackouts discussed at the beginning of this chapter. Although the precise limits would still be a subject of some dispute, the southern extent of the "arctic" ionosphere was generally taken to lie at about 57° geomagnetic latitude and was defined confidently only for the Northern Hemisphere. According to this boundary, all ionospheric data in Canada were considered "arctic data." In terms of the area encompassed within the zone of the arctic ionosphere, members of RPL were quick to point out that the ionosphere above Canada was the most representatively "arctic" in the world.[32] (See figure 3.11.)

Northern Communications

Beyond the laboratory, that linking of radio disruptions and Northern geophysics drove a profound ellipsis and a remarkable inversion. Whereas Meek's vision had been to trace out the detailed mechanisms of ionospheric disturbances, often through lengthy and complex chains of physical causes, others used those efforts and their products to link Northern nature and problematic communications more immediately. A 1958 history of the Defence Research Board, the organization that oversaw RPL's work starting in the late 1940s, drew the connections by way of national identity: "Radio communications in Canada are bedeviled by difficulties which are unique." In the next sentence, it went on to list not communications difficulties at all, but the geophysical phenomena implicated in them: "The North Magnetic Pole lies wholly within Canadian territory; the North Geomagnetic Pole on the north-western edge of Greenland is closer to Canada than to any nation, and the phenomenon of the aurora borealis, or northern lights, has an adverse effect on radio communications throughout most of the Dominion."[33] The ellipsis here—citing the markers of northernness *as* unique communications difficulties—is significant. It suggests the way communications disruptions could point directly and unproblematically to a characteristically Northern natural order. And in doing so it pointed to how, rather than having the features of the North explain communications disruptions, the disruptions themselves might be used to define what it meant to be Northern.

That idea of technological failures *as* natural phenomena resonated broadly. It had already appeared in the work of Harold Innis and other prominent commentators. In the early 1950s, Innis, the economic historian whose "staples theory" had dominated postwar understandings of Canadian history, had turned his attention to communications, elaborating a macro-historical view of political and economic control in which the character and extent of communications technologies broadly conceived—railroads, waterways, trade routes—determined the consolidation and the fate of states and empires. Innis' musings were set against a vision of a nation held together tenuously by communications technologies opposed and thwarted by nature. That idea traded on a long-standing view in Canadian history, articulated by Northrop Frye and others, of a North that defined Canada while threatening its survival. On that view, communications had allowed Canada to emerge as a nation in spite of its geography. For Innis, that understanding was precisely backward: "The present Dominion was created not in spite of geography, but because of it."[34] Canada's Northern geography had shaped its communications systems, on Innis' view, and those systems, in turn, had given the present nation its identity and character, its economic and political relations. But this did not stop unreliable communications from threatening the country. In 1952, Innis charged that the situation of Northern radio, captured in the map that opened this chapter, "threatened Canadian national life."[35]

Innis' theories were enormously influential throughout the 1950s, when he held a number of prominent academic positions within Canada and internationally, and they deeply informed the work of his former student and colleague Marshall McLuhan. The work of RPL and others during the 1950s traced the same connections as that influential view, but in reverse. Whereas Innis claimed that the North shaped communications, and through it the nation, here it was communications, and particularly communications failures, that shaped and defined the very idea and meaning of "North" and of Canada. The work of the Canadian Broadcasting Corporation illustrates the extent of that inversion. In 1958, the CBC established its Northern Service. Through broadcasts in English, French, Eskimo, Dene, Cree, Chipewayan, and Loucheux, the Northern Service sought "to provide a broadcasting that will meet the special needs of all the people of the North—Indian, Eskimo, Métis and others—and give them a sense of identity with the rest of Canada as well as with their own northern community."[36] In studying the possibilities for Northern broadcasting, the CBC had turned to RPL to help with studies of short-wave broadcasting to the Yukon and the Mackenzie District of the Northwest Territories, the largest audiences for the future Northern Service.[37] In the next ten years, the frequency assignments for most of the Service's radio stations were determined through a combination of field reports and ionographic analyses conducted at RPL.[38] The

CBC was well aware of the influence of Northern geophysics on radio transmissions in the region. "Although radio would appear to be an easy answer to the problem of communications among the scattered population of the North," one report explained, "this area is subject to some of the worst broadcasting conditions in the world. This is the result of atmospheric interference caused by the north magnetic pole, which lies within the territory." When it came to defining the region that their broadcasts were meant to serve, however, the CBC inverted those connections. Its officials conceded that their "North" had no precise boundaries.[39] Rather than impose what seemed the arbitrary and unhelpful borders of political jurisdiction, officials of the CBC turned to "the inherent nature of radio transmissions" as a guide.[40] "The Northern Region," they explained, "extends north to the Pole and south *to an imaginary line that would include those listeners who do not receive a consistent and adequate broadcast signal from CBC network stations or private stations located 'outside'* [the region]. By this definition, it covers almost two million square miles and has, by the 1961 census, a population of about 80,000."[41] On this view, the image that opened this chapter was as much a definition of the North as it was an illustration of the threats the region and the nation faced.

Those connections were even more explicit in the work of Louis-Edmond Hamelin, professor of geography at Quebec City's Laval University. Initially trained in classics and economics, Hamelin spent time in the late 1940s at the Scott Polar Research Institute in Cambridge and in Paris (where he obtained a doctorate in geomorphology from the Sorbonne in 1951). Throughout the late 1950s, his interests in the Northern polar regions culminated in the establishment of the field of nordology, the study of the cold regions of Northern latitudes. At the core of his discipline, Hamelin placed the concept of "nordicity": the "state or quality of northern-ness or being north."[42] Unsatisfied with a qualitative definition, he moved to quantify northernness through a "nordicity index," a quantitative measure based on a number of criteria that would help distinguish a nation's "true" nordicity from false claims to that title. Meteorology, geography, topography, and population studies all helped to define his ambitious index, as did the undifferentiated criterion of "communications." In a move that resonated strongly with the premises of the Radio Physics Laboratory's work, Hamelin asserted that (subject to minimum latitude and isothermal requirements) the poorer a region's communications were, the more properly Northern it was.[43] Combining these features into a comprehensive system over the next ten years, Hamelin determined that archetypal

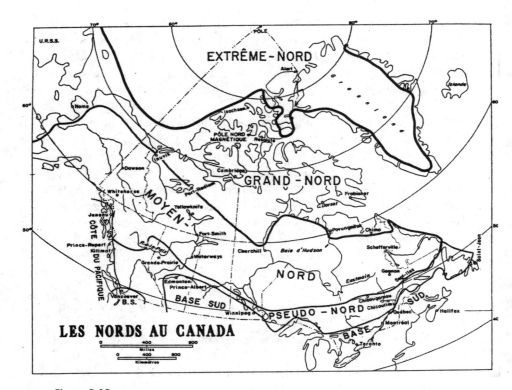

Figure 3.12
Louis-Edmond Hamelin's "nordicity index," applied to Canada. Hamelin used the index to determine the extent of Canada's "nordic" territory, using indices that drew on the unreliability of radio communications. Later versions of the index would replace the broad category of "communications" with transportation and surface access. Source: Louis-Edmond Hamelin, "Essai de Régionalisation du Nord Canadien," *North* 11 (1964). Image courtesy of Indigenous and Northern Affairs Canada.

Northern nations such as Iceland, Scandinavia, and the Soviet Union enjoyed much less "severe nordicity than that of Canada." His calculations allowed Canada's Northern area to be estimated at approximately 70 percent of the country. (See figure 3.12.) "Given this extent," he concluded, "the territory is no less than a national feature."[44] For Hamelin, the oppositional relationship between communications and northernness in the 1950s was self-evident and presumptive. When he revisited his scheme in the late 1970s, by which time satellite communications had begun extending reliable communications throughout Canada, the communications criterion had disappeared from his index altogether.

Conclusion

Seen through the eyes of Louis-Edmond Hamelin and his contemporaries, the specific drama that played itself out through Northern radio expressed Canada's historical relationship to the region and its position in the Cold War. But beneath those contingencies, many observers suggested, lay a more elemental relationship between nature and technology. It had begun speculatively, as a general association between a Northern natural order of geomagnetic disturbances and auroral storms, on one hand, and a topography of communications disruptions centered on the North Atlantic on the other. But by the late 1950s it had taken on a far more specific form: a set of often visual arguments for Canada's archetypal northernness and its essential relationship to unreliable communications. Those arguments, initially running along causal chains from violent geophysics to radio communications, ultimately created the conditions for their own inversion. Instead of the features of the Northern natural order explaining communications difficulties, radio failures came to define the very idea of North around which the nation's character and identity were being built. They enacted simultaneously the nation's northernness, its hostile nature, and the threats against it during the Cold War. They formed part of a web of exhaustive and intricate links that tracked back and forth, between nature and specific groups of machines, passing through a set of scientific images whose interpretation was made possible by a transformed way of seeing.

The attempts to make the high-latitude ionogram legible ultimately redefined the working objects of the discipline. The sporadic phenomena that characterized records from the field stations of the Canadian North had been curiosities after the war, excluded from the exhaustive correlations and predictive machinery of radio communications. The initial attempts to make those phenomena readable still privileged the well-defined traces found in temperate regions. And they ultimately introduced more uncertainty, as unqualified and inexperienced field workers struggled to find the traces in the seeming chaos of actual ionograms. Ultimately, reading the high-latitude ionogram meant abandoning the paradigm of moderate latitudes and crafting instead a reading regime suited specifically to the peculiarities of high-Northern atmospheric phenomena. The f-plot, with its emphasis on mere witnessing over immediate analysis, ultimately allowed Canadian and other high-latitude researchers to link the visual evidence of ionospheric disturbances to Northern geophysics and radio disruptions in ways that had previously been impossible. The instruments, practices, and concepts created a "regime of perceptibility" in which the connections between the North,

its characteristic natural phenomena, and radio behavior became both possible and self-evident.[45] By including those data within the quantitative
arena, it allowed researchers to define a polar ionosphere to match the
machinic order they had laid out a decade before.

In this way, the laboratory's work fed back into broader cultural narratives that had supported its research in the first place. The very phenomena
that characterized the high-latitude ionogram were used to naturalize disrupted communications. DRTE's histories of ionospheric disturbances furnished powerful narratives about the North and communications that were
inverted in the wider culture of Cold War Canada. Where the work of the
lab aimed to start with the Northern natural order as a way of explaining
why communications were fractured, others began with fractured communications as a way of defining what it meant to be Northern.

We end this chapter where we began, then: with the image of radio
waves entering the Cold War North. But we have benefitted from this tour
through the anxieties that surrounded ionograms in the first decade of
the Cold War. Those anxieties illustrate how the attempts to carve out a
place of influence, even power, during the period were waged not only on
the scale of Canada's vulnerable Northern expanses, but also at the micro
level of scientific practices, field station offices, and the images that occupied them. By the late 1950s, the new methods for reading ionograms had
made it possible to articulate, scientifically and authoritatively, the details
of a hostile Northern natural order whose defining phenomena opposed
reliable shortwave communications and the larger projects of national completion and geopolitical control based on them. In cutting off the Canadian
North in this way, that natural order of DRTE's investigations made the
region more vulnerable; in making the region more vulnerable, it intensified
the hostility of the preeminent hostile environment of the period. By that
time, however, new machinic orders were emerging. Under them, the turbulent natural order of the high-Northern ionosphere would itself become
the most hostile of all the Cold War's environments.

4 Hostile Environments and Cold War Machines

On July 8, 1962, at 22:00:09 local time, a hydrogen bomb exploded in the upper atmosphere 400 kilometers above the mid-Pacific islands of Johnston Atoll. The tiny US territory had served as a federal bird sanctuary in the 1920s. During World War II, it had been converted into a refueling base for submarines and trans-Pacific bombers, including the *Enola Gay*—the plane that dropped the atomic bomb on Hiroshima. On that July night in 1962, the nuclear connection had come full circle. The detonation, code-named Starfish Prime, was part of a series of high-altitude US nuclear weapons tests hastily initiated after the Soviet Union announced the previous August that it would end its three-year moratorium on nuclear testing. The larger series of detonations aimed to answer long-standing questions that had emerged from the Pacific Proving Grounds over the previous decade: Why did atmospheric nuclear detonations generate auroras precisely halfway around the world? Why did their damaging electromagnetic pulses vary with altitude? What was the spatial distribution and intensity of the communications blackouts they caused? Starfish Prime and its allied tests aimed to investigate those phenomena using warheads more powerful than ever before; the remote Johnston Atoll had been chosen as the launching point so that the detonations would not flashblind civilians 1,400 kilometers away in Hawaii or in the more distant Marshall Islands.

The tests were also designed to investigate less subtle questions; more brute effects that extended beyond concerns about the immediate properties of nuclear weapons themselves. The electromagnetic pulse from the blast, more powerful than expected, knocked out nearly 300 streetlights on Oahu.[1] Its effects spread skyward, generating an enormous belt of intense artificial radiation that spread 30° to each side of Earth's magnetic equator, damaging several orbiting US satellites. Additional rockets streamed into the radiation cloud from the Hawaiian Islands to collect data.

Alongside an interest in contamination rates, fallout patterns, radiation levels, and dispersion mechanisms, the detonations were designed to simulate the new, hostile operating environments that faced Cold War machines. Up until the late 1950s, the Arctic had formed the archetypal hostile environment of the Cold War—remote, sparsely populated, inadequately mapped, and a potential future battlefield, its extreme conditions made it a privileged proving ground for the universal soldiers and machines envisioned by Cold Warriors.[2] By the late 1950s, that interest had shifted to two new environments—the oceans and outer space—and the machines that would occupy them: submarines hiding and operating under the immense pressure of the deep sea; missiles passing through a hail of cosmic rays and micrometeorites in the near vacuum of space; or satellites operating in the radiative remains of a thermonuclear detonation. Overlaid constantly with real and potential nuclearity, these hostile environments became the focus for concerns about the relationship between the proper functioning of Cold War machines and the newly salient physical media through which they moved.[3]

As part of that larger reorientation, DRTE had begun recasting the high-latitude ionosphere from remote and exceptional natural order to the preeminent hostile environment of the Cold War, a region that acutely concentrated the dangers and problems of machines operating in near space. Although their research program on short-wave radio disruptions continued, their work increasingly envisioned a different machinic order comprising satellites, ballistic missiles, and the radar systems designed to detect and surveil them. The centerpiece in that recasting would be a project known internally as S-27, a joint venture with NASA that would produce the first satellite designed and built outside the United States or the Soviet Union. Conceived by its engineers as a miniature laboratory, S-27 would investigate the near-space environment and its threats to weapons and surveillance technologies; as a machine functioning in that environment, and scheduled to be launched just weeks after Starfish Prime had irradiated its future orbit, the satellite was the site of intense anxiety about the reliability of machines, and particularly electronics, operating in exceedingly hostile conditions.

This chapter explores the place of satellite technology in the ionosphere's transformation from natural order to hostile environment. It does so first at the level of broad postwar geographic conceptions and later through the design, testing and construction of S-27 itself. The hostile environments of the Cold War emerged from an exceptional geography of threat created by World War II operations and elaborated through wartime social and environmental science. One of the major threats these environments posed was the wide-scale failure of advanced military technologies. Alongside its other ambitions, early space research was an attempt to investigate and document the new hostile environment of near space, including the ionosphere, in order to better understand the dangers it posed to machines. Previous space histories have repeatedly dealt with how the archetypal technologies of space—rockets, satellites, probes—functioned as symbols of national prowess.[4] As DRTE's work shows, however, questions of identity were also worked out through the detailed practices that produced those objects. Just as the design and construction of the S-27 satellite functioned as highly visible signs of national achievement, its structure, material details, even the symbols that limned its circuit diagrams, formed sites where its designers tied together the distinctiveness of the high-latitude upper atmosphere and the problem of failing machines.

Despite what media releases claimed, the S-27 project was arguably never designed to improve short-wave communications. Instead, it formed part of a broader practice of using the threatened failure of machines to define natural, environmental, and even geopolitical hostility during the period.[5] In addressing the problem of electronic failure, DRTE's engineers drew on an emerging military philosophy of hyper-reliability, generated by widespread concerns about electronics operating in punishing environments, and deployed that philosophy behind the attempt to generate ionograms from space. This chapter therefore forms a hinge between DRTE's decade-long attempts to use the ionogram as an instrument for articulating the natural and machinic orders of high-frequency radio on one hand, and the future use of the records as increasingly symbolic objects—pointing beyond themselves to engineering cultures, machinic reliability, and peaceful uses of outer space—on the other.

Machines and Media

World War II was organized around specific "theaters" of war—Europe, the Pacific, the Mediterranean, and North Africa. With military forces spread into almost every part of the world, one pressing challenge was how to gather and organize detailed knowledge on actual and potential zones of conflict. Leading into the war, that information had generally been assimilated to one of two dominant geographies: a continental approach that took continents as the prime organizing category, rooted in the Asian subcontinent, and a European colonial approach, which carved the world into a series of colonial possessions. Both of these, however, were poorly suited to a conflict that itself reconfigured global geopolitics. What emerged gradually alongside these other understandings, rather than displacing them, was a new system of world regions. Rather than continents or empires, regions represented more narrowly bounded units, defined by the ostensibly coherent "cultures" or "societies" that occupied them.[6] Overwhelmingly, they were characterized by their potential for threat to wartime operations, machines, and personnel. As a result, military planners created a set of institutions based in the human and natural sciences that would provide the geographical knowledge needed for worldwide operations. In October 1943, the Office of Field Services was created under the direction of Vannevar Bush and Karl Taylor Compton—the same people who would later push for the centrality of technology to the Cold War—in order to explore the specific difficulties posed by transportation, communication, climate, insects, and distance in Pacific operations. The wartime Ethnogeographic Board, made up largely of anthropologists, published *Survival on Land and Sea*, which exhaustively detailed differences in flora and fauna across the various regions and suggested universal techniques for dealing with hostile human populations from the tropics to the Arctic. Those activities created not only a spatial understanding of logistics, troop dispositions, and violence, but also a geography of hostility and multiple threat anchored in the crude but effective categories of the Tropics, the Desert, and the Arctic—areas whose hostility as potential operating environments was overdetermined by geopolitics, climate, and terrain.[7] The postwar period normalized this exceptional geography, creating the initial "hostile environments" of the Cold War—"terrain types which might, it was believed, be scenes of conflict under the umbrella of planetary struggles with fascism and then communism."[8]

Like human subjects, machines helped define the hostility of these environments. The Arctic had been the site of investigations into equipment failure under extreme conditions even before the dramatic human acclimatization trials that characterized winter warfare simulations. RPL's detailed postwar connections between Northern geophysics and radio blackouts had added another dimension to the perceived antagonism between environment and machines, particularly the electronic devices at the heart of monitoring, communications, and surveillance activities. In 1948, Lloyd Berkner—a radio engineer on Richard Byrd's first Antarctic Expedition (which Frank Davies had also joined) and the future chief architect of the International Geophysical Year (IGY)—addressed precisely those concerns in a lecture to the US National Academy of Science on the design of electronic equipment for cold-weather use. Berkner, who understood the links between science and national security better than most, explained that the

Arctic formed one of three "arctic" regions that included the cold regions of the atmosphere experienced during very-high-altitude flight (from about 5 miles to the warmer regions of the ionosphere, his own area of expertise). Aircraft operations in the polar regions faced problems almost identical to arctic work: the solder that fused electronic components together deteriorated at low temperatures, mechanical insulation failed, and lubricants lost their most important properties. But Berkner focused particular attention on the malfunction of electronic components. Temperature sensitive capacitors, for instance, "can cause the complete failure of the apparatus of which they are a part."[9] Strobes or gated circuits, such as the ones in an ionosonde, would often stop working altogether if their circuits underwent wide temperature changes. Berkner worried about how these failings of electronic parts would spill over and cascade into larger problems as they spread through electronic devices in extreme conditions.

In temperate climates, these malfunctions were mere inconveniences; in the world's cold regions they foretold human failure and even death. Mixing the malfunctioning of machines with the language of catastrophe, and echoing the wider anxiety of Cold War nuclear gaming at the RAND Corporation and elsewhere, Berkner emphasized the need to plan "for any eventuality, any unforeseen event that may be of consequence."[10] His cautions underlined the role of machinic failure in defining environmental hostility and motivated a broader, urgent search for a set of principles, materials, and concepts that would allow the full range of technologies—from delicate scientific instruments to brute-force atomic weapons—to operate reliably across the world's "arctic" environments.

In the late 1950s, particularly with the development of ICBMs, the focus on the Arctic gave way to concerns about what contemporaries called the "new world environments" of the oceans and near space. The shift inspired a large-scale expansion and transformation in the earth sciences that helped define the new imaginative geographies of the Cold War.[11] Two general interests drove that expansion. The first was a concern about signals, particularly the propagation of sound and radio waves involved in detection, evasion, and surveillance in the oceans and near space. The second was an attempt to catalog and understand the conditions that threatened the operation of machines in those environments.[12] The threat from submarine-launched nuclear missiles in 1952, for instance, spawned a wide-ranging research program into the physical characteristics of oceans—temperature gradients (thermoclines) and topographical features that provided barriers to sonar and potential hiding places for submarines; depth and pressure studies to aid with hull designs and stealth operation; acoustical research to determine underwater sonic signatures, shaft and blade speeds, cavitation rates, and engine cycling characteristics of specific vessels; sound propagation studies for discrimination between submarines and whales or large schools of fish—all linked together through a network of low-frequency listening stations for wide-scale surveillance and detection.[13] The second main interest was in these environments as spectacular sites to probe the limits and vulnerabilities of vessels. Underwater nuclear tests such as Operation Wigwam (a 30-kiloton deep-water detonation off the California coast) were designed to determine the susceptibility of submarines to deep-water blasts. The shallower Baker Test spectacularly incinerated Bikini atoll, but the "maelstrom" of radioactive debris it created was directed at "target ships" in its coral lagoon, decommissioned submarines drawn from the American fleet, and captured German and Japanese ships intended to inform the design of survivable military hardware.[14]

Research into the upper atmosphere and near space was also focused on these twin questions of signals and survivability. The International Geophysical Year was, for all its other ambitions, an attempt to gather information on near space as an *operating environment* for space vehicles and a detection site for ICBMs. In the late 1950s, anti-ICBM systems depended on infrared radars to detect warheads as they passed through the lower atmosphere, about two minutes before impact, allowing the missile and its payload to fly undisturbed for about 30 minutes beforehand. Fears of a Soviet surprise attack in 1954 prompted President Dwight D. Eisenhower to create the Technological Capabilities Panel of the Science Advisory Committee, headed by MIT president James R. Killian. Its report, known as the Surprise Attack Study, urgently recommended expanding and accelerating research on ICBM detection, particularly radar detection systems that gave warnings of an attack within minutes, while warheads were still in the upper atmosphere and the exosphere.[15] The panel additionally recommended launching a scientific satellite during the IGY—the future Vanguard project—to establish the principle of freedom of space (foreclosing future objections that satellites passing over Soviet territory violated that nation's sovereignty).

All three of the IGY's chief architects—James Van Allen, Sydney Chapman, and Lloyd Berkner—were experts in upper atmospheric research. Chapman, future president of the IGY Special Committee, was a specialist in magnetic storms and aurora; Van Allen had been directly involved in rocketry and *in situ* experiments. Berkner, a well-respected ionosphericist, understood and sought to exploit the intersections between international geophysical research, foreign policy, and national security.[16] The more general

IGY program on space research, in which Berkner took a special interest, was directed precisely toward surveillance and tracking problems of near space and questions about the ability of machines to operate and survive in that environment. Cosmic-ray studies and the discovery of the Van Allen Belts suggested that near space simulated the effects of a nuclear detonation and created fears of prolonged radiation exposure to machines and humans. Near space also presented environmental threats in particularly acute form—micrometeorites and sand blasting from space dust; ruptured seals and evaporated lubricants from the high vacuum; wild temperature fluctuations that threatened electronic instrumentation; radiation and decay; the impossibility of maintenance and repair; the shielding of human bodies. Keeping even modest amounts of equipment working unattended, for months at a time, required good knowledge of what a 1958 RAND Corporation publication called "severe environmental conditions." But even those effects were poorly quantified. The RAND study emphasized again and again the unknowns—the lack of precise knowledge about hazards. To illustrate, it calculated that a spherical space vehicle 3 feet across, with skin 1 millimeter thick, could expect to be punctured during a time frame from 3 months to 170 years.[17]

As advanced technologies and weapons systems pushed into these new environments, many worried that existing ways of designing and building machines were wholly inadequate.[18] Asked in 1945 to imagine a future war, the aerodynamicist Theodore von Karman had envisioned "a war machine in the proper sense of the word, consisting of technical devices only," capable of operating anywhere in the world, and exploiting developments in supersonic flight, pilotless aircraft, all-weather flying, perfected navigation and communication, and imperviousness to environmental conditions.[19] At the core of von Karman's ideal was a widely held vision of hyper-reliability and failure-proof technology that faced serious challenges early on. Already in 1949, one study estimated that Navy electronic equipment was malfunctioning 70 percent of the time, with critical surveillance systems such as radar and sonar making up the bulk of the failures. Electronic malfunctions during the Korean War were implicated in the failure of tactical missions and in aircraft disasters. On June 14, 1952, a US Marine Corps Panther jet exploded after a faulty fuse accidentally armed and detonated one of its 260-pound fragmentation bombs shortly after takeoff. Three months later, on September 10, a radar malfunction sent six of the same jets, returning from a mission in dense fog near Sariwon, into a mountainside. Over the course of the Korean conflict, the threat that electronic failure posed, not only to limited missions but also to the broader strategic aims of the Cold War itself, led some to claim that the whole success of the Cold War defense effort was bound up with the problem of reliability.[20]

That disjuncture between vision and reality in Cold War technologies created what contemporaries called the "Reliability Crisis"—an episode that launched the most urgent, wide-ranging, and comprehensive examination of electronic reliability ever mounted, reaching its height around 1957 and 1958 and coinciding with the IGY. It spawned competing understandings of reliability that would play an important role in the central missions of the space age, and in the satellite project mounted by DRTE.

The S-27 Project

The S-27 project began in the midst of this combined anxiety about hostile environments and complex, failing machines. Starting in 1957 and coinciding with the IGY, DRTE's newly created Communications Laboratory began reorganizing its research activities to focus on the "disturbed ionosphere," specifically the radar aurora caused by the disturbed conditions that followed the ejection of particles from the sun.[21] The intense radiation produced by those conditions provided an ideal testing ground for sensitive electronic equipment.[22] With the shift in emphasis from Northern territorial warfare to defense against ballistic missiles, the lab's reorientation recast the ionosphere from an erratic reflecting layer for short-wave signals to a hostile operating environment for missiles and spacecraft. The issue of survivability would very soon occupy the members of the lab directly, but their main focus in 1957 was on another problem.

Projects like the one at the Communications Laboratory treated the ionosphere as a distorting lens, a medium through which the ultra-high-frequency (UHF) radar signals had to pass in order to observe ICBMs and other space vehicles lying beyond. The main concern of the laboratory was that the enormous influx of solar particles would generate radar aurora, generating spurious signals or auroral clutter in UHF radars and interfering with detection and tracking of missiles coming over the polar regions.[23] As the laboratory's director, John Chapman, explained in August 1958, "The purpose of this program is to tell us how much trouble AICBM [Anti-Intercontinental Ballistic Missile] radars will be in."[24] In pursuit of that goal, the laboratory had begun using various proxies for ICBM echoes. Radio signals passing through the ionosphere from the sun, planets, and radio stars would give information on the inaccuracies of radar measurements and insight into the radar characteristics of missiles.[25] But Chapman and his colleagues also recognized "an urgent need for new knowledge of the environment in that space a few hundred to a few thousand miles above the planet, in order to appreciate what physical principles we may be able to call on in devising a defence."[26]

As the Soviets had already recognized, satellites served at least two important functions in investigating that environment.[27] As distant objects under surveillance from Earth, placed into orbit by modified ballistic missiles, satellites could serve as proxies for other passive radar targets, such as nuclear warheads. By comparing the discrepancies between optical and radio tracking, researchers could determine radar inaccuracies introduced by the ionosphere—believed to be up to 80 seconds of arc in 1956—and reduce them with good predictions of atmospheric conditions.[28] The Communications Laboratory had begun doing this in late 1957, tracking Sputnik in order to "catalogue the flying ironmongery and discover if this is a useful way to measure accuracy of radars."[29] Using careful measurements of the satellite's orbit, they could also determine the shape and strength of Earth's gravitational field. But as machines operating in near space, satellites also provided direct knowledge of the operating environment. The early Vanguard satellites were designed to measure temperature and micrometeorite strikes, for example.[30] Leonid Sedov, Chairman of the Interagency Commission on Interplanetary Travel, explained in February 1956 that the purpose of the Soviet "laboratory" that would eventually become Sputnik was to measure details of the environment around the satellite—pressure, temperature, air resistance, quantity, and speed of meteors, ultraviolet radiation, and cosmic rays.[31] Sedov and his colleagues on the commission were detached from most of the eventual details of the project developed by Sergey Korolev, one of the heads of Soviet ballistic missile development. But Korolev's specifications for the satellite were also built around this investigation of the environment in the vicinity of the spacecraft: the sphere was kept highly polished to aid optical tracking; the shape was changed from cone to a metallic sphere, to determine atmospheric density in its path; the radio pulses sent out by the satellite were carefully designed to encode information about temperature and pressure, indicating whether the shell had been punctured by a meteorite.[32]

The idea of a satellite specifically dedicated to ionospheric research had been part of the earliest satellite proposals. The IGY had set up 170 ionospheric stations conducting hourly soundings every day for 18 months.[33] But because radio waves emitted from the ground would only be reflected up to the point of maximum density, the regions of declining density above roughly 400 kilometers were effectively invisible to ground-based sounders. Soviet scientists had proposed a satellite to get past these limitations as early as 1955, and the calls only intensified after the discovery of the Van Allen Belts. In July 1958, toward the end of the Communication Laboratory's reorientation toward ICBM research, the US National Academy of Science's Space Science Board under Lloyd Berkner, eager to promote space science after the formation of NASA, requested proposals for satellite experiments to be launched by the United States. James Scott, a former Royal Canadian Air Force squadron leader and an original member of OIC/6 who had taken over as DRTE Superintendent when Frank Davies moved to DRB headquarters, approached the head of the Radio Physics Laboratory, Colin Hines, about the possibility of developing instrumentation for a third-party satellite. Hines, wary of the coordination involved in using another agency's satellite body, suggested that his lab might think about taking on the project in a year. Eager to see if it could be undertaken sooner, Scott turned instead to the Electronics Laboratory, led by Eldon Warren and involved in bottom-side sounding experiments. Warren embraced the idea and started working on a proposal for the National Academy of Science.

In September, Warren and several colleagues traveled to Boulder, Colorado to meet with the Space Science Board's Working Group on Satellite Ionosphere Measurement. Feeling that the proposal had a better chance of acceptance if it was understood as a project in advanced engineering, they shied away from a simple fixed-frequency sounder of the kind that had been launched on rockets up to that point and instead proposed to replicate a more complicated sweep-frequency sounder capable of producing complete ionograms identical in format to the ones produced from ground-based field stations. They calculated that, by closely spacing the adjacent frequencies, a relatively detailed picture of the ionosphere with a resolution of about 10 kilometers could be produced. Since the data from the topside were considered crucial in determining the behavior of the ionosphere as a whole, the extension of the technique seemed promising. By creating a formal symmetry between top-side and bottom-side records, the proposal would allow researchers to more easily reconcile data from the two regions. (See figure 4.1.) Finally, as the laboratory emphasized in an aide memoire to the Minister of Defence, by synchronizing the satellite transmitter with the receiver of the ground sounders, the symmetry made possible trans-ionospheric transmission studies useful for ICBM radar research.[34] By late 1958, the project had taken its place in a three-part space program at DRTE—alongside rocket soundings for the IGY and radar studies of the moon and planets—dedicated to the ICBM threat.[35]

In the next few months, the proposal evolved to emphasize the threat to machines from the environment of near space, particularly above Canada.

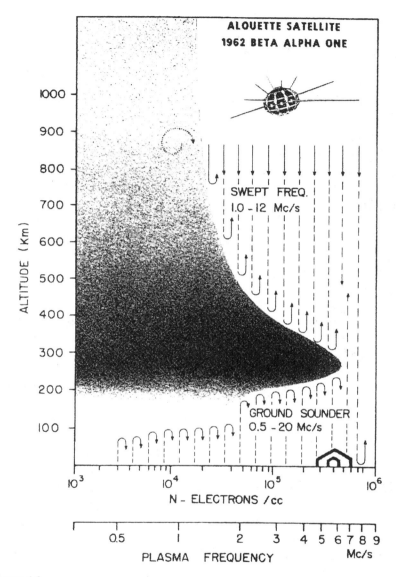

Figure 4.1

A schematic representation of the technological symmetry envisioned by Warren and Chapman. (S-27 would be renamed Alouette after its launch in 1962.) The shaded region represents the electron density of the ionosphere as derived from ionograms. Source: A. R. Molozzi, "Instrumentation of the Ionospheric Sounder Satellite 1962 Beta Alpha (Alouette)," in *Space Research IV* (Proceedings of the Fourth International Space Symposium), ed. P. Muller (North-Holland, 1964). Copyright Elsevier (1964).

In September, Warren, Scott, and two colleagues proposed the project at a separate meeting at Cornell convened by Henry Booker, Berkner's former collaborator in radio propagation research and the wartime head of Britain's Telecommunications Research Establishment. The DRTE proposal emerged as the preferred option, but the Academy canceled its program shortly afterward.[36] Scott, undaunted, proposed the project to the US Advanced Research Projects Agency (ARPA) at a meeting at the Pentagon in mid November. He detailed the nucleus of the proposal: "A satellite should be placed in a high inclination orbit instrumented with a sweep-frequency 2–20 Mc/s iono-sonde to measure over the period of a year exospheric electron density-height distributions especially at high latitudes over the auroral zone."[37] Since this was an international proposal, officials at ARPA recommended that it should be formally submitted to the newly created NASA. In looking for support from the Defence Research Board, Scott recast Canada's relationship to the high-latitude ionosphere, the basis for the laboratory's work on communications problems since the 1940s, as a threat to *space technologies*: "The Canadian ionosphere and the exosphere above it, due to high geomagnetic latitudes and the auroral zone, present special complex environmental problems of ionization and radiation to high altitude rockets and satellites. Canada thus has a unique opportunity for research in this area."[38]

The DRB, itself immersed in developing Canada's strategic interests and nuclear thinking independent of the United States, wove the satellite project into wider and more fantastical visions of space activities late in 1958.[39] In envisioning a potential space program, the overseeing board speculated about the probabilities of one-hour transatlantic flights on ballistic trajectories and manned vehicles for military control of the moon, planets, and other regions of space.[40] Like so many other space initiatives, it cast those projects as part of a nationalist narrative that both emphasized the importance of these activities for geopolitics, economic development, and science, but also connected them to the peculiar natural order that the nation occupied.[41] It portrayed Canada as a possible future base for space launch facilities, made faintly plausible by the suggestion that radiation trapped by Earth's magnetic field was less intense near the North Magnetic Pole, making the large land masses of the far North an attractive site for manned vehicle launches. Because of these effects, the United States and the Soviet Union were interested in gaining more information. Echoing almost exactly the late-war suggestions to leverage high-latitude ionospheric research, turning it into a source of status and power, the report suggested space science as a way for Canada and other countries to carve out a sphere of influence in space activities. Launch facilities were likely out of the question: the capabilities and resources needed for multi-stage rockets would almost certainly restrict them to nations with ballistic missile programs, which Canada and other non-superpower nations lacked. But a shortage of expertise in experiments and instrumentation for space research, and for radiation studies in near space in particular, meant that the United States and the Soviet Union were willing to launch satellites or space probes designed and built by other countries. For basic science, as well as applied concerns about communications and ICBM tracking, the report noted that satellite experiments would be instrumental.[42]

NASA officials were impressed with the proposal when Scott approached them in late December, but skeptical about the technical challenges. A sweep-frequency sounder would require nearly continual power generation during sounding; and in order to sweep across the frequency range, it would need to be outfitted with 150-foot antennas somehow contained inside a much smaller satellite body and deployed in orbit. NASA accepted the proposal but asked the National Bureau of Standards' Central Radio Propagation Laboratory to examine it for scientific merits. In June 1959, CRPL confirmed NASA's worries that the project might be too ambitious and recommended that it be developed as a second-generation experiment, alongside a fixed-frequency sounder that CRPL would develop in conjunction with the Airborne Instruments Laboratory. The two teams—American and Canadian—would form a joint working group to exchange information and expertise. The American satellite would study ionospheric profiles along the 75W meridian, while the DRTE project would explore the environment of near space over high latitudes, with the United States providing the launch vehicle, tracking, and telemetry.[43]

S-27 had been completely integrated into the project of investigating the environment of near space by December 1959. Over the course of the next year, the satellite gradually evolved into a miniature "laboratory."[44] The initial proposal had called for a small payload—a baseball-sized package of electronics that would essentially miniaturize a ground-based sounder and replicate its most important characteristics—sweep rate, pulse repetition, and frequency range. In the summer of 1960, the Canadian National Research Council proposed to include particle counters to measure electron, proton, and alpha-particle fluxes at the satellite, giving information about the intensity of the radiation on the horns of the outer Van Allen Belt at high latitudes and their variation with upper-atmospheric disturbances such as solar flares and auroras.[45] Two other additions followed: a cosmic noise experiment, which merely involved extending the lower limit of the ionospheric receiver, and a VLF recorder, which would study the propagation of very low-frequency radiation believed to follow magnetic field lines.

Mounted at the height of the reliability crisis, the S-27 project was shot through from its beginnings with an intense concern for reliability. Failure plagued early space projects. In the late 1950s, most spacecraft had a lifetime of only a few months, complicated by the difficulties of both reaching and maintaining orbit.[46] Sputnik had lasted precisely three months before its orbit decayed and it reentered the atmosphere. In designing the Soviet satellite, Sergei Korolev emphasized that one of the guiding principles would have to be "maximum simplicity and reliability." He was seconded by Sedov who again emphasized simplicity to ensure reliable operation, and signaled it in the satellite's eventual designation—Simple Satellite No. 1 (PS-1).[47] Even when spacecraft achieved orbit and stayed there, failures in even moderately complex electronics were legion. NASA had shied away from producing a sweep-frequency sonde such as S-27 precisely because of the challenges its complexity posed. Andrew Molozzi, S-27's Satellite Controller, would later explain that the one-year planned life of the satellite meant that the design team paid special attention to reliability: "The approach taken placed particular emphasis on the integrated design of the circuitry, that is, components in the circuit rather than on individual component parts alone."[48] Molozzi's cryptic assertion would be restated, in varying forms, over and over—an emphasis on parts in the circuit rather than on individual parts alone—until it achieved the status of engineering mythology. It was an explicit attempt to distinguish the stance of the Electronics Laboratory against a powerful alternative. Nowhere in the voluminous documentation of the S-27 project can we find a coherent and detailed elaboration of the philosophy that guided the satellite's design; but in the actions and reports of those involved in its execution, we can identify the technical embodiments of a specific understanding of reliability, the practical dimensions of a philosophy prescribing the conditions for trusting electronics, particularly space electronics, at the height of the Cold War.

Trusting Machines

The earliest organized responses to the Reliability Crisis of the mid-1950s focused on the most obvious and spectacular faults in electronic equipment, known as "catastrophic failures." Aptly termed given what was at stake, catastrophic failures represented incidents where electronic equipment failed to operate altogether rather than simply malfunctioning. They seemed to be everywhere around critical electronic equipment in the early

1950s, and they forcefully argued for locating the causes of failure in individual electronic parts. Electronic *assemblies* and *equipment* might be failing, but the physical sites of electronic failures were the fundamental electronic entities that composed the larger configurations—the blown-out vacuum tube, the shorted capacitor, the smoking resistor.

These parts represented the atomic constituents of electronics, "an item which cannot be disassembled without destroying its identity."[49] As the focus of concern, their failure bred a kind of atomism in relation to reliability, supported by an emerging calculus developed within aerospace engineering but transposed onto electronics. That calculus was most fully elaborated by Robert Lusser at Redstone Arsenal, a wartime chemical weapons plant and rocket facility that became one of the centers of American ballistic missile research and space engineering in the 1950s. Lusser, formerly a German aviator, an aircraft engineer for Messerschmitt and Henkel, and a co-designer of the V1, was brought to the United States as part of Operation Paperclip along with his rival, Wernher von Braun. At Redstone, Lusser concentrated on the theoretical study of reliability in complex systems, developing Lusser's Law—an atomized calculus of reliability that stated that at its simplest level (what Lusser and others called "inherent reliability") the reliability of a system was given by multiplying the reliabilities of its individual components.[50] On this understanding, maximizing reliability meant incorporating extremely robust individual elements into the simplest possible designs. Lusser would use the law to claim that von Braun's boasts of reaching the moon and planets would founder on the rocks of complexity.

This atomistic view—anxious about catastrophic failure, locating reliability in fundamental components, disdainful of complexity and distrustful of design—dominated the approach to military technology throughout the mid-1950s and was reflected at the highest levels of the ballistic missile program. Six months before von Braun's ABMA team was transferred from Redstone to the newly formed NASA, J. M. Colby, chief of Army Ordnance Missile Command (which included ABMA, Redstone, JPL, and White Sands), attacked the growing move toward system engineering and the complex designs that resulted, complaining that there had "been more crimes committed under the concept of system engineering than any other crusade. The system is only as good as its weakest link. The system is only as good as its weakest component."[51]

Inside and outside the military, in fields from communications to atomic weapons and space vehicles, this atomism was taken up by those interested in developing systems quickly and in making existing systems more robust.

And throughout the 1950s, similarly broad-based criticisms of the approach surfaced. They first of all questioned the ability of such an approach to produce reliability at all. In his attempts to develop a system of reliable military command communications in the early 1960s, Paul Baran at the RAND Corporation would make precisely this point in avoiding the traditional practice of telephone companies who "tried to increase the reliability of the system as a whole by making each component as reliable as possible."[52] Baran instead favored using more modest components, designing the system to be insensitive to the vagaries of its individual parts through the use of redundant units, for example, or by rerouting information around failed nodes in the network. For Baran, the survivability of the communications system rested in its ability to adapt and compensate for the shortcomings of its individual elements.[53]

This practical criticism was often coupled to an epistemic concern. Despite the proliferation of advanced statistical models for calculating reliability, critics held that knowledge about the reliability of a system could not adequately be derived from knowledge about the reliability of its individual parts. In 1961, the US House Armed Services Committee articulated precisely this opinion in its consideration of guided missiles. Members of the HASC argued that "testing of the components of a nuclear missile system separately was not adequate to rule out the possibility that some feature of the way they were combined would prevent the missile from working."[54] Knowledge about reliability, like other aspects of system performance, did not scale up neatly.

These criticisms formed the basis of a holistic understanding of reliability that emerged alongside the atomistic approach in the late 1950s, particularly in the wake of Sputnik. In Canada, it was particularly strong in the Air Force's Air Material Command, which worried that although the atomistic approach might ensure the reliability of some standard military electronics, it could not secure critical electronic systems operating in the "new world environments" of the Cold War. Guidance systems and proximity fuses for missiles, electronic systems for supersonic aircraft, even instrumentation for satellites—these systems, critics argued, operated under conditions so punishing and hostile that they demanded a level of reliability surpassing anything the atomists could produce. Rather than focus on catastrophic failures "due to their dramatic obviousness," the holists focused on "degradation failures"—the less dramatic but much more pervasive malfunctions caused by the gradual degradation of parts.[55]

In a series of documents produced throughout the 1950s, Chester Soucy, a leading member of the Air Material Command and the Canadian military's most vocal champion of the holist cause, captured the practical and epistemic dimensions of the approach. Soucy first of all argued that the practice of focusing on improved parts as "a panacea for equipment failures" could never ensure reliability to the required levels. In a document produced as the S-27 proposal was taking shape, in October 1958, and presented to the Institute of Radio Engineers, he laid out in graphic simplicity the inadequacy of the atomist program. (See figure 4.2.) When the atomists retorted that a return to simplicity was the answer, Soucy countered: "our major military problem is not, as it is sometimes urged, that of making equipment simpler—desirable though this obviously is—but, more realistically, it is that of making necessarily complex equipment more reliable."[56] Soucy backed up his criticism by pointing to an epistemological flaw in the atomist program. In his numerous letters to RCAF officials, in his discussions with colleagues at DRTE's Electronics Lab, in his talks to the Institute of Radio Engineers (IRE) and other organizations, Soucy railed against the very definition of system reliability employed by atomists: "Such an oversimplified representation interprets equipment reliability principally in terms of the reliability of parts, and it neglects the large measure of control of the reliability performance level due to the variations between different operating-maintenance environments such as airborne, fixed-ground, shipborne, missile-borne, etc."[57]

Against the reduction of the atomists, Soucy and the holists focused on both parts and circuits; they emphasized the conceptual elements that contributed to reliability; they rejected the idea of the complex as the enemy of the reliable; and they emphasized design as the decisive element. The bulk of the reliability problem stemmed not from weak components, but from design flaws. In place of the atomist calculus, Soucy offered an alternative definition of "inherent reliability": $R_I = R_i \times K_{design} \times K_{manufacture}$. Here the atomist calculus entered as only a single term (R_i) given by the product of individual part reliabilities. Assuming a high quality of construction (which both atomists and dualists agreed was lacking), the design-modifying factor (K_{design}) furnished the critical term. It was a function of various design principles (often adapted from mechanical engineering) that "make the equipment impervious to the physical and the operator-maintainer environment."[58] The use of "Standard Proven Parts" was one of these principles, and Soucy sanctioned the rigorous testing, research and development that established the robustness of individual devices. But to Soucy's mind, and to the minds of engineers throughout the Department of National Defence, reliability would only be achieved "by standardizing on, not only proven parts, but on proven circuits and modules."[59] These "Standard Proven Circuit

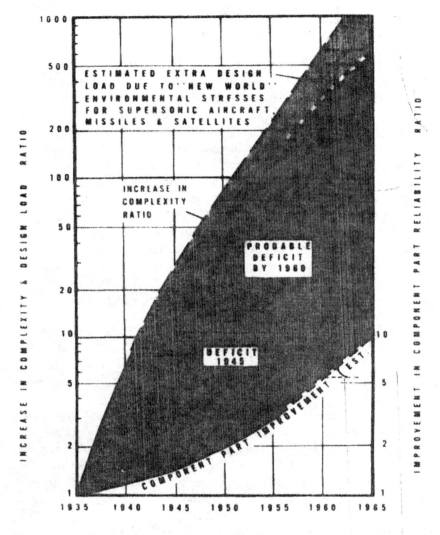

TREND TOWARDS INCREASING RELIABILITY DEFICIT
IN WEAPONS SYSTEMS *UNLESS* IMPROVEMENT IS
MADE IN SYSTEM AND EQUIPMENT DESIGN

Figure 4.2
The graphic argument for the inadequacy of the atomist program. Soucy's graph shows the increasing complexity of designs needed for "new world" electronics such as missiles and satellites (upper curve), and the improvement in electronic parts reliability (the lower curve). The shaded area between them represents the reliability "gap" or deficit that lay at the heart of the "reliability crisis." Source: Chester I. Soucy, "The 'Project MATURE' Concept of the Design and Production Requirements for Reliable and Maintainable Military Electronics Equipment," LAC, Record Group 24, accession 1983–1984/049, volume 1664, file 1950–123-7, part 2. © Government of Canada. Reproduced with the permission of Library and Archives Canada (2016).

Blocks" would incorporate additional design principles and practices—circuit tolerance to the environment and to part degradations over time, redundancy, safety margins, de-rating—that ran against the atomist desire for simplicity, but would guide even inexperienced engineers in the process of creating reliable electronics.

Determining whether these principles had been properly applied could be determined "only by tests to destruction of component parts and assemblies."[60] Under the holist approach, this form of testing would expose weaknesses in the structure of the electronics that might otherwise lie hidden in circuit diagrams or in the complexes of silicon and solder that were their material embodiments, only to emerge during the rigors of missile launch or supersonic flight. By deploying this testing regime against the circuit as well as the individual part, equipment could be made to "age" more quickly within the controlled environment of the engineering laboratory. But just as crucially, the entire practice and epistemology of electronics could be more profitably focused on the holistic nature of reliability, revalorizing the "skills of the circuit and system designers" disparaged by Lusser, Colby, and the atomists.

Throughout the mid-1950s, the Electronics Laboratory of DRTE formed one of the most prominent sites for this expanded, holistic approach to electronic reliability, particularly around questions of design. Formed at the outbreak of the Korean War, the initial purpose of the EL was to evaluate the suitability of new components for military applications, like guided missiles and microwave radar, and to occasionally redesign existing telecommunications equipment to take advantage of new developments in electronic parts. By 1952, when the EL obtained its own building on Montreal Road, it formed part of an expansive research complex for military electronics, adjacent to Canadian Signals Research and Development Establishment, heavily involved in everything from materials research to medical electronics.[61] In early 1954, to better focus its efforts, Frank Davies authorized the creation of a Components Section, to be supervised by Frank Simpson. Subdivided into Solid-State Research, Magnetic Research, and Component Techniques, Simpson's team worked on making individual parts more reliable. They coordinated with wider efforts in standardization, oversaw industrial contracts for electronic parts and furnished the secretariat for the Electronic Components Development Committee (ECDC), the organizational center of the most widespread reliability program for electronic parts in Canada and the link between the EL and similar efforts in the United Kingdom and in the United States.

One group of components, however, escaped the scrutiny of Simpson's team. Transistors instead fell under a second section of the Electronics

Laboratory, also created in 1954, and led by Norman Moody. Moody had started his career in the British Telecommunications Research Establishment (TRE), where he had worked for six years on radar circuits and had experienced the reliability problem at first hand. From TRE, he had moved on to the radio and television industry, and then to a four-year position as head of the Electrical Branch at the Chalk River nuclear laboratories northwest of Ottawa, before coming to DRTE in early 1952. His interests in the months after his arrival at the Electronics Laboratory focused on transistor triggers, and he gradually gathered around him a group of engineers and technicians dedicated solely to investigating the emerging field of transistors and their circuits.

Smaller, lighter, and physically more rugged than vacuum tubes, transistors held out the hope of smaller, faster, more robust electronics. In the early years of their development, however, transistors were problematic devices. Early point-contact transistors had limited frequency performance, were noisy, had little power, and were vulnerable to power surges, radiation, and high temperatures. Even the germanium junction transistors introduced in 1951 and commonly used through the mid-1950s were highly temperature-dependent (failing to work altogether at about 75°C) and suffered from relatively large leakage currents, which caused problems for the switching circuits required by digital electronics.[62]

Simpson's section concentrated on improving the material characteristics of transistors. But the reliability of the devices also depended on different circuit techniques. Moody's group therefore worked on carefully linking transistor parameters (like base-emitter voltages or input impedance) to circuit parameters (such as current at the collector) to develop an understanding of how the properties of the individual part varied with changes in the larger electrical environment of the circuit. From here, Moody's team devised transistor-specific *design* solutions that would both keep the devices within tight electrical parameters and make the circuits employing them relatively insensitive to variations in the devices' performance. In designing a digital computer for use in DRTE, for example, Moody and another engineer, David Florida, used the investigations of the Transistor Group to produce a "family of circuits that thrives on heavy load currents, often to the limit set by transistor destruction." Singling out one circuit from their family of configurations, Moody and Florida went on to tout its properties. "This circuit," they claimed, "*though relatively complex in itself, yields an excellent performance in the face of wide transistor and component variations.*"[63] Insensitivity to these variations in the logic circuits of the computer was instrumental for computational accuracy (the computer would be used to reduce data from S-27) and the work of the Transistor Group hinged on this avoidance of even minor malfunctions through circuit design. Using Thompson's symbol in all its design work, Moody's section had grown by the mid-1950s to include several subdivisions—Transistor Measurements and Radio Frequency Receivers, Digital Techniques, and Advanced Circuit Research. The foundational work that went on there was recognized in the new title given to the section: Basic Circuits. (See figure 4.3.)

This type of rigorous investigation of both robust parts and robust circuits within the same laboratory was rare in the Canadian defense establishment. The standardization of electronic *parts* in the Armed Services had long been the responsibility of a series of organizations, of which the Canadian Military Electronic Standards Agency (CAMESA) was the culmination. Questions of *design* were organizationally separate, despite criticisms from within CAMESA itself, and fell under the jurisdiction of the Design Authorities of the Armed Services, responsible for guiding the proper application of parts in military electronic equipment. Although nominally sovereign in the area of design, the hands of the Design Authorities were effectively tied in the mid-1950s when the move toward standard equipment within NATO forced the Canadian military to increasingly adopt predominantly American designs for their equipment, despite the sometimes more punishing operating conditions experienced in the Canadian North. In a defense establishment where research into robust parts was organizationally separate from investigations of robust circuits, where continental standardization often stripped the Design Authorities of their authority to design, the Electronics Laboratory formed a rare organizational space where the evaluation of parts *and* circuits could profitably come together, and where the often political act of design, executed in symbols specific to the Canadian military, could be carried out in relative freedom.

In their respective capacities, Simpson and Moody brought the considerable abilities and resources of Components and Basic Circuits to bear on a number of projects undertaken for the Armed Services. They had collaborated with their colleagues at Canadian Signals Research and Development Establishment (CSRDE) on guided missiles and on the development of a five-mile radio set; they had worked on proximity fuses and electronic countermeasures for the Air Force. And they had additionally come together on the largest project the EL undertook during the 1950s: the Doppler navigation radar for the CF-105 Avro Arrow, a Canadian-made supersonic interceptor whose controversial cancellation in February 1959 would fracture the nation's aerospace community.[64] For six years the Doppler Section of the Electronics Laboratory, standing alongside Components and Basic Circuits but drawing heavily on the material and human resources of these divisions, worked on reconciling the functional requirements of radar with

ELECTRONICS LABORATORY

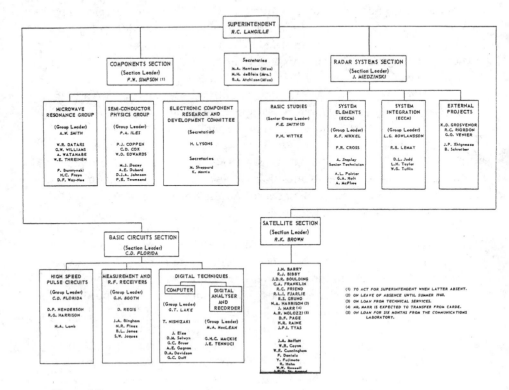

Figure 4.3

The organizational chart for the Electronics Laboratory. The chart shows DRTE's organization at the start of construction on S-27 (January 1, 1960). Note the presence of both the Basic Circuits Section and the Components Section, an organizational embodiment of the holist approach to electronic reliability. Source: Defence Research Board, "DRTE Scientific and Administrative Organization," DRTE Publication No. 1037, January 1960, Library and Archives Canada, Record Group 24, volume 9, file DRBS 100–22/0, part 2. © Government of Canada. Reproduced with the permission of Library and Archives Canada (2016).

the robustness required for supersonic flight. It enlisted Simpson's team, with its connections to CAMESA and its involvement in the atomist program, to carefully screen individual components. The Doppler Section called on members of Moody's transistor group to transistorize the design. For more than a year, guided by the reliability demands of Chester Soucy and the Air Material Command, a group of technicians and a design team led by Moody's newest protégé, Colin Franklin, struggled to combine electronic stability and temperature compensation in the components with the necessary radiation characteristics for the radar. That is, understanding the dire consequences of malfunctions in navigation systems during supersonic flight, Franklin and his team sought to reconcile radar fidelity with electronic reliability. Using the non-standard representation for the transistor in their designs, they went on to evaluate competing formulations over prescribed temperatures ranging from −55°C to +70°C (almost the precise range later used for the S-27 systems) before arriving at what they believed was an optimal version in late 1958. The Doppler project would ultimately be canceled before full flight testing, but the members of the design team would carry their philosophy of reliability with them when, in early 1959, with most of their engineering and technical staff intact, they formed the core of the Space Instrumentation Section to begin work on S-27. (See figure 4.3.)

Material Culture and Hostile Environments

The satellite project drew heavily on surrounding resources. Special electronic hardware poured into the lab from the Department of National Defence. Materials and personnel that would help secure the proper operation of the spacecraft were drawn from every corner of DRTE, stretching thin its human and material resources. Frank Davies, who took back his duties at the head of DRTE in 1960, complained frequently about the "albatross" that had been hung around his neck.[65] Engineers and technicians skilled in semiconductor electronics and digital circuits, their mechanical counterparts from the machine shop, ionosonde technicians from the Arctic division of the Radio Physics Lab, and radio engineers from the newly formed Communications Lab came together in early 1959 to form the Satellite Section. By April it was positioned alongside the other major sections of the lab—Basic Circuits, Radio Warfare, and Components—in DRTE's organizational chart. (See figure 4.3.)

Three main groups composed the Satellite Section. A Development group was in charge of assembling components and performing equipment checks. A Mechanical Design team headed by Chief Mechanical Engineer

John Mar would develop the satellite body. Mar had completed his engineering studies at the University of British Columbia in the early 1950s. In 1955, he had designed a 16-millimeter movie camera with an exceptional large-aperture lens for auroral investigations at the Defence Research Northern Laboratory in Churchill, Manitoba. He then moved on to aircraft engineering at DeHavilland, when the Electronics Laboratory recruited him to work on the mechanical design of S-27.[66] Finally, a System Design team led by Chief Electrical Engineer Colin Franklin was charged with the electronic design and testing of the satellite. Franklin had recently completed PhD work at Imperial College London in semiconductor circuits and operation before joining the Electronics Laboratory in 1957. His first major duty had been to lead the engineering team in charge of transistorizing the Doppler navigation radar for the Air Force. He was joined by several other Imperial College PhDs, including Andrew Molozzi and the lead engineers of the Doppler project. (RPL physicists would describe the S-27 satellite as a "space-based radar.") For more than a year, starting in early January 1959, Franklin and Mar struggled with the general engineering aspects of the satellite before finally delivering the general specifications and schematics for the spacecraft in the spring of 1960.

The major concerns about the satellite's reliability came out of its focus on the high-latitude ionosphere.[67] In designing S-27, the lower limit of the frequency range had to be considered carefully to make sure that records produced above the Canadian North would contain enough information. The relatively stable ionospheric conditions in most parts of the world meant that the lower limit was generally located around 12 megacycles per second (Mc/s). In polar regions, however, and in the ionosphere above the Canadian North in particular—with the influence of the Van Allen belts and the North Magnetic Pole—the limit could be as low as 2.8 Mc/s.[68] For the Systems Design group, this lower frequency meant three things: cosmic noise could possibly interfere with the sounder echo; the satellite would require antennas at least 15 meters long, longer than any previously used on a satellite; and extra power would be needed to overcome the spatial losses.[69] The length of the antennas represented a serious mechanical problem. The extra power required by the electronics raised deep concerns about longevity and performance. In order for S-27 to gather sufficient data to distinguish diurnal and seasonal variations, the mission would have to last at least one year. A near polar orbit—very similar to the orbit spy satellites would occupy—had been chosen so that the rate of precession of the satellite would be just over two revolutions per year. This would take S-27 just south of Alert (the northernmost inhabited point in Canada and an important military base) but would cover all other parts of the country. A more equatorial orbit would sacrifice data on the Canadian North; a more northerly one would extend the necessary operating period beyond one year and possibly jeopardize the quantity of data gathered.

The Satellite Section spent the spring and summer of 1960 attacking these problems.[70] Rocket firings to measure cosmic noise levels in the winter of 1959 had gone ahead and failed. A faulty antenna deployment system on both Javelin flights (in December 1959 and January 1960) had caused the antennas to become detached from the rocket. Finally, in mid May, a cosmic noise recorder was flown on a Transit IIA satellite, and gave the required design information. But the problems of antenna deployment that had jeopardized the rocket experiments also threatened the satellite. Since the resources of DRTE in both personnel and material were stretched thin, the antenna system was contracted out to DeHavilland, an aircraft manufacturer in Toronto, which would develop sounder antennas for the range of frequencies. By the summer of 1960, the project was still in the design stage and the method of deployment still uncertain.

Although the original agreement with NASA had specified that Canada would provide only the instrumentation for a satellite, the special antennas required that DRTE also develop the satellite body.[71] Mirroring newly adopted protocols for military equipment, the Satellite Section would undertake staggered development of four satellite models—one development model, one prototype, and two flight models.[72] These would be developed sequentially to determine the general feasibility of the system design and to allow trial modifications without jeopardizing the integrity of the final satellite. According to Mar's calculations, the satellite body would need to be an oblate spheroid 42 inches in diameter and 34 inches in height. It would form a clamshell, constructed around two thrust tubes with attached mounting platforms for the antennas and electronics. (See figure 4.4.) The shape and dimensions were a compromise between three main requirements: accessibility of the electronic packages for testing, temperature control, and steady illumination of the six 480 solar cells (obtained from Lockheed Missile Company) that covered the satellite body and provided power to the storage batteries and electronics.[73] The problem of incorporating a 150-foot antenna into a 42-inch satellite was solved in principle by drawing on wartime work at the National Research Council of Canada. A strip of spring steel 0.004 inch thick and 4 inches wide was passed through a guide and then heated to take the form of a tube or cylinder with an overlapping seam. It was then carefully opened while being wound back onto a drum (on the model of a carpenter's rule), all the while retaining its tendency to form a tube if unconstrained. By attaching a motor to the drums, the flat strip could be extended through a guide sleeve that allowed the

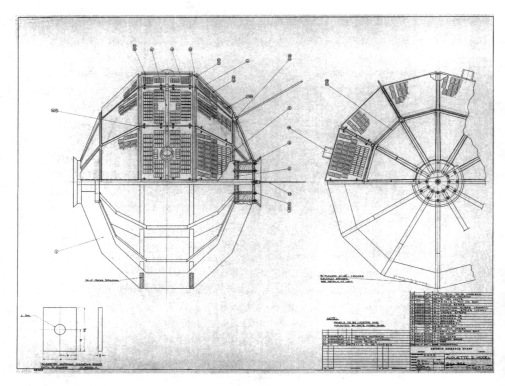

Figure 4.4

A drawing showing the design of the satellite body specified by Mar for Alouette: an oblate spheroid 42×34 inches. The circular element on the center line of the left image is one of the four ports for the antennas. Source: Library and Archives Canada, Department of Communications collection. © Government of Canada. Reproduced with the permission of Library and Archives Canada (2016).

"memory" of the steel to dominate and form a tubular antenna.[74] The antenna system would eventually be used on nearly all subsequent Canadian and American satellites and spacecraft throughout the next two decades.[75] But ensuring the mechanical reliability of the extension mechanism would occupy Mar's efforts for much of the project.

In order to help protect the satellite electronics, Mar specified that the outer shell of the satellite would be made of polished aluminum. So as not to complicate the already formidable electronic system further, Mar had decided to utilize a passive temperature control system. Together with the optically coated glass of the solar cells and a combination of black and white enamel paints, the metal would give the satellite body the proper absorption-emission ratio for solar radiation. By selecting materials with the proper optical qualities, Mar hoped to satisfy two opposing temperature requirements: that the solar cells operate as near as possible to 0°C for peak efficiency and that the inner instrumentation package and storage batteries be kept at about +27°C. Other satellites, among them Explorer V and Explorer VI, had achieved this using extendible solar arrays, which introduced mechanical complexity. For the sake of reliability, Mar had opted for an approximate spherical shape for the satellite, minimizing complexity even at the expense of electronic efficiency.[76] Loose thermal coupling between the shell and the satellite's interior and a so-called Kropshot blanket (an insulating sheet of aluminized Mylar produced by the sewing facilities of the Defence Research Chemical Laboratory's Materials Section) would hold the instrumentation bay between −7°C and +45°C, fluctuating around the optimal operating temperature for the electronics: +27°C.[77]

By the time Mar delivered these thermal calculations in the Summer of 1960, Franklin's team had completed the overall electronics design for the satellite. The instrumentation package for S-27 would be divided into four subsystems—sounder, secondary experiments, command and telemetry, and power supply. The last of these ultimately supported all the others, channeling power from the solar cells to the storage batteries (six in total to compensate for high failure rates in other satellites) and to the satellite's command and telemetry systems.[78] These formed the nerve center of the satellite. Chapman and Warren had decided against storing data on board the spacecraft. Instead two telemetry transmitters sent housekeeping information—battery levels, payload and shell temperatures, instrumentation voltages, and currents—along with raw ionospheric sounding data to the ground stations, which recorded them on magnetic tape. Four small turnstile antennas on the upper hemisphere of the spacecraft received very-high-frequency (VHF) radio commands, turning the satellite on or off and allowing satellite controllers in Ottawa to optimize the operation of the satellite by switching in redundant units in case of failure, or shutting down secondary experiments in order to preserve S-27's primary mission of ionospheric sounding.[79]

Franklin's team focused their attention overwhelmingly on the sounder system, and particularly its transmitter. The sounder formed the scientific centerpiece of the S-27 instrumentation. Because the operating parameters of solid-state components like transistors and diodes varied with temperature fluctuations, and because maintaining these parameters in the transmitter was important for the formal production of the ionogram, securing the reliability of the transmitter blocks became the major electrical problem of the spacecraft. From the point of view of the electronics team, the greatest threat came from the wide temperature fluctuations experienced as the satellite passed from full sunlight into Earth's shadow and back again. Beyond Mar's structural design, there was little they could do to protect the instrumentation from micro-meteorite strikes or cosmic radiation, which would gradually degrade the components. But the temperature variations affected the power and frequency characteristics on which the quality of ionograms depended, potentially calling into question the validity and usefulness of the primary mission. In dealing with one component of the transmitter, for instance, Franklin had initially favored using a motor-driven capacitor of the kind used in ground sounders, since it would work across a wide range of temperatures within normal atmospheric pressures. But the near vacuum of space created serious problems. "Unfortunately," Franklin explained, "the vapor pressures of available lubricants limit bearing lifetimes, in high vacua, to less than 1,000 hours. A pressurized container is therefore required for the motor, tuning capacitor and associated gear trains. In addition, there are problems of mechanical reliability and ruggedness, and satisfactory feed-thru connectors for the walls of the container."[80] Ground tests had also revealed a jitter in the mechanical sweep that created range ambiguities on the order of ±100 miles in the resulting ionograms.[81] The change to a solid-state device solved the problem of atmospheric pressure, but made the transmitter seriously susceptible to temperature fluctuations below 0°C, affecting the linearity of the frequency sweep. The more linear the sweep, the less dependent it would be on changes in transistor parameters due to temperature. "The problem," Franklin noted, "proved less difficult than expected." He then went on to expose the dual concern which had pervaded the Electronic Laboratory's approach to reliability: "This has been partly due to the availability of improved silicon transistors and the use of low leakage Mylar capacitors, but also to the development of improved circuit techniques."[82]

The combined emphasis on parts and design practices that characterized the work of Basic Circuits would run through Franklin's work on the amplifier, the most difficult of all the electronic problems his team faced. The design team had originally considered a continuous-wave amplifier that would give new fine-structure detail in the resulting ionograms. In the fall of 1960, however, they realized that it produced too much interference between transmitter and receiver, and they abandoned it in favor of a more traditional pulsed sounder. The pulsed amplifier required at least 100 watts of peak power in order to overcome the spatial losses at frequencies used to sound Northern geomagnetic latitudes. In attempting to meet these power requirements, the Systems Group had two main options: a high power vacuum-tube amplifier that would easily distinguish signal from noise, but would also dissipate much more power, operate at higher voltages and lower reliability, and require warm-up times for the tube heaters; or a low-power transistor amplifier that would be more efficient and reliable, but would sacrifice the signal to noise margins and lead to a possible loss of ionospheric data at long ranges.

The group now broke up to determine which version should be included in the final payload. John Barry and Harold Raine took on the tube version, while Franklin and Page investigated the solid-state amplifier. Despite their advantages in the area of power, vacuum tubes were notoriously unreliable as components and, in the spring and summer of 1961, the causes of their unreliability were still poorly understood. Barry and Raine set out to subject candidates to a punishing testing regime set up for the EL's recent work on navigational radar. Vacuum tubes initially considered for the transmitter amplifier, for example, passed through an extensive screening process, identical to military electronics, before moving on to the Electronics Laboratory. The centerpiece of the evaluation process, however, was a program of "severe environmental and lifetime testing" that involved testing-to-destruction. They placed two glass tubes on heater cycling, a suspected cause of failure. A constant current source powered one tube and a constant voltage source the other. The testing continued through the fall of 1960 and in December, Raine reported on the status of the tube tests: "To date these tubes have been turned on more than 18,000 times with adequate heating and cooling times allowed. Although a test of two tubes cannot be statistically valid, this early evidence indicates that heater cycling is not likely to be an important limitation and that a constant voltage source can be used without a 'wetting' current during transmitter 'off' periods."[83] During the spring of 1961, Raine and Page had also started two additional tubes on high-power pulsed operation, cycling them continuously for 2,000 hours. Through March and into the summer, the tube heater cycling continued alongside the pulse tests. In March, it had reached 42,000 cycles with no failures, and by June it had been discontinued after 50,000 cycles without issue. "Exhaustive" tube checks had been carried out "to detect any possible deterioration in tube characteristics."[84] The set-up was then modified to produce faster cooling and two more tubes were started on the regime. "The transmitter is being designed," they asserted, "to operate at conservative temperatures while those tubes on life test are deliberately operated at their maximum rated temperature."[85] They incorporated two tubes into different amplifier stages and began testing them under high-power operation. In July they had undergone almost 2,000 hours of operation.[86]

By the summer of 1961, Raine and Page were satisfied with their results. The life tests had shown that tubes could be reliably used in the amplifier stage and that the system design ensured this reliability for ranges far beyond operating conditions. In the meantime, Franklin had been working on a transistor model. New transistors being used in NASA's fixed frequency sounder (S-48) had just become available. From the information at hand, Franklin believed the resulting circuit could be capable of supplying the required power with greater efficiency and reliability than the vacuum-tube configuration. But reports from the designers of S-48 suggested that the transistors had suffered failures even when operating well within their ratings.[87] In January, new transistors had arrived and been incorporated into the breadboard model of the amplifier. The development model of the electronics, complete except for the command equipment and telemetry transmitter, had been checked out on the bench. "Severe stability problems" resulted when the amplifier reached levels of 5–6 watts, Franklin reported. "Furthermore," he continued, "the difficulties were common to the three available types of high frequency transistor tested."[88] After careful redesign, Franklin described the test results:

> Throughout the last quarter, two converters—one regulated the other unregulated—have been continuously cycled through a normal load, partial overload, and short circuit sequence every 35 seconds. No change was observed in the performance of either of these converters and it therefore seems likely that the basic design incorporates adequate protection against overloads and damaging switching transients. . . . The protection problem has been emphasized since the design of an efficient and highly reliable converter is one of the more difficult problems in solid state electronics.[89]

The amplifier showed slight fluctuations in output over the frequency range, and small changes in the high-frequency cut-off at the extremes of the temperature range, but was considered to be within the limits for producing acceptable data. In the spring of 1961, Franklin and Page exposed the amplifier to continuous half-hour intervals of operation at −50°C and +75°C, with 25 percent fluctuations in supply voltage, before running it for 12 hours at +75°C with no change seen in performance. "It should be emphasized," Franklin cautioned, "that acceptance of this amplifier is dependent on satisfactory completion of life tests at elevated temperatures." The tests were encouraging enough for him to conclude that the flight models would most likely include a pair of 100-watt solid-state amplifiers.[90]

The temperature was now increased to 85°C (again, well above the expected outer shell temperature of the spacecraft). The transmitter breadboard model had undergone 1,000 hours of operation at this temperature, modifications had been made in the mounting of some of the transformers after vibration tests, and Franklin now laid out the program to come: "Accelerated life tests at 110°C and 135°C will be carried out during the next quarter."[91] The results, Brown reported, were encouraging: "soundings have been made from the ground and the ionograms transmitted by telemetry link, received and recorded on tape and played back from the tape." The flight models now went into urgent production with an extra twenty workers put on the program, most of them dedicated to construction and testing of component circuits.[92]

Unlike the later tests for the entire payload, these were *not* simulations. Those came later. It had been more than a year since John Mar had specified in October 1960 that the payload temperature was expected to fluctuate between −7°C and +7°C in 66 percent sun orbit, staying constant around 45°C in full sun orbit, and that line voltages could be held to ±5 percent.[93] According to Franklin, the tests were instead intended to create an environment even more hostile than that of near space. The aim was to "expose design weaknesses that you would want to correct anyway even though [the circuit] would apparently pass all the tests at the strictly limited operating range." With some additional engineering, these design weaknesses could be fixed, meaning that transistor parameters would have to go "wildly" out of range before affecting the ionogram.[94] Only after the surviving circuit designs had been incorporated into the spacecraft did the simulations begin. The flight models underwent spin and balance tests at DeHavilland in October and November, were shipped back to Ottawa within 48 hours for more electrical checks, then sent to Goddard Space Flight Center for vibration tests. From here, they returned to the Electronics Lab for further electrical checks and then passed to a specially built thermal vacuum chamber at the Canadian Armament Research and Development Establishment (CARDE).

On October 24, Mar circulated his test plan to simulate the "thermal space environment" using the CARDE thermal vacuum chamber. After placement of the satellite on its rotating stand and instrumentation of the outer skin for temperature measurements, the satellite models would undergo "100% Sun Simulation." The vacuum chamber temperature would be lowered under refrigeration. As the chamber pressure reached 0.0001 mm Hg, the rotating stand would be turned on and lamps would illuminate the satellite. (See figure 4.5.) The spacecraft temperatures would then be

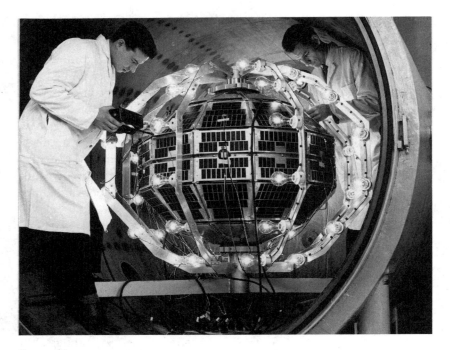

Figure 4.5
Environmental testing at the Canadian Armament Research and Development
Establishment. The satellite body was placed on a rotating stand to simulate its rotation in orbit. Controlled lighting was used to simulate heating and cooling of the
spacecraft inside the vacuum chamber. Source: Library and Archives Canada, Department of Manpower and Immigration collection. © Government of Canada. Reproduced with the permission of Library and Archives Canada (2016).

measured every 15 minutes to obtain the heating transient. The illumination continued until the satellite reached a steady state temperature both
inside and out, and its temperatures would again be recorded. At this point
of highest temperature, Mar instructed, "run electronics to simulate maximum turn-on duty cycle. Record all S/C [spacecraft] temperatures every 15
minutes until new steady state is reached. Record electronic payload functions."[95] Once the satellite had again reached steady state, temperatures
were recorded and the test was terminated.

The satellite then entered "66% Sun Simulation" to mimic the condition
of the spacecraft during the coldest period of its precession, initiated without
break from the previous test. The arced lamps would now start a cycle of
69 minutes "on" and 47 minutes "off" until a steady-state temperature was

again achieved. "When S/C [spacecraft] reaches steady-state temperature fluctuation, record 3 complete cycles by reading all S/C temperatures every 10 to 15 minutes. During this stage, electronics will be operated to simulate duty cycle corresponding to minimum sun operation."[96]

From here, the spacecraft would undergo a simulated ascent into orbit. Less than three minutes after launch, the Thor-Agena rocket would jettison its first (Thor) stage and shed the payload shroud that protected S-27 during launch. Since the satellite would not yet be spinning, it would be vulnerable to solar and aerodynamic heating during the roughly six-minute second-stage burn. Because of this, officials decided it should be launched at night under the protection of Earth's shadow. In order to simulate heating during ascent, the arc lamps would now be turned off and the satellite allowed to cool to 85°F. Once the satellite reached this temperature, rotation was stopped and the arc lamps were turned on for 50 minutes while temperatures were recorded at 10-minute intervals. Lastly, the satellite entered its low temperature soak: "With chamber vacuum purged, cool S/C down to about −20°F or to whatever limiting low temperature chamber is capable of cooling S/C to. At minimum temperature, test and record electronic payload functions. Terminate test and remove S/C from chamber."[97] After the tests in late February 1962, each model was shipped to the Pacific Missile Range, the US Navy's center for development and testing of its Atlas ICBMs, for booster and separation system tests throughout the spring and the early summer.

Late in the summer of 1962, about six months behind schedule, the flight model was moved to Vandenberg Air Force Base (the launch head for the Pacific Missile Range), which specialized in placing satellites in polar orbit. There it underwent final checkout and was transferred to the launch pad. The launch time (11:30 p.m.–1:30 a.m., September 28–29) was set so as to allow the satellite as many soundings as possible on its first few orbits without full exposure to the sun, in case it suffered from a malfunction or catastrophic failure during initial heating. Each successive orbit would expose it incrementally to more sunlight while technical staff checked its status on the ground. On July 9, about eleven weeks before the scheduled launch, the US conducted its Starfish Prime test. In the subsequent days and weeks, the radiation from the detonation damaged several orbiting American satellites. In early September, with the scheduled launch of S-27 only weeks away, Frank Davies wrote anxiously to his superior in the Department of National Defence. After briefly describing the intensity and extent of the radiation from the July detonation, he revealed that the orbit of S-27 would intersect the belt, offering one of the first opportunities to study the prolonged effects of an upper-atmospheric detonation, but also placing the satellite directly in the radioactive remains of the explosion.

Rather than postpone the launch, Franklin and Mar advised it should proceed as scheduled. On September 29, 1962, at 23:09 Pacific Daylight Time, S-27, aboard a Thor-Agena B rocket, was launched into a polar orbit. A US Navy tracking ship in the Indian Ocean had been assigned to monitor the satellite's antenna extension, but in the moments after launch the ship suffered equipment and operator trouble and lost contact. A ground tracking station in Johannesburg eventually picked up the satellite signal and confirmed the antennas had worked as designed.[98] Two hours after launch, the satellite was turned on from its control center in Ottawa and produced its first ionogram around midday on September 29. (See figure 4.6.)

Worried about its imminent failure, scientists scrambled to gather the greatest possible amount of data from the approximately 1,100 ionograms per day that S-27 (now renamed Alouette) began transmitting to eleven telemetry stations around the world.[99] Two weeks later in Ottawa, at a joint meeting of the American and Canadian national committees of the Union Radio-Scientifique Internationale and the Professional Groups of the IRE, the Canadian team presented the first results of the satellite, eventually

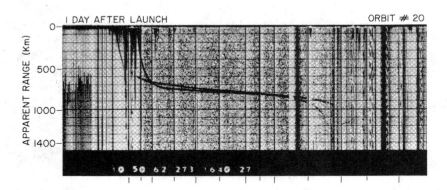

Figure 4.6

One of the first topside ionograms ever produced, this record was generated by Alouette I one day after its launch. As in bottomside ionograms, the two overlapping traces represent different propagation modes. The spikes extending downward around 1.5 MHz (plasma spikes) are unique to these topside records and are produced by energy stored in the plasma around the satellite. © 2016 IEEE. Reprinted, with permission, from John E. Jackson and E. S. Warren, "Objectives, History, and Principal Achievements of the Topside Sounder and ISIS Programs," *Proceedings of the IEEE* 57, no. 6 (1969): 861–865.

published in *Nature* early in the next year.[100] They began with a series of unexpected phenomena not observed on bottom-side ionograms: a plasma spike created as the satellite's immersion in the ionospheric plasma triggered resonant oscillations in the medium; a cyclotron spike where the energy from the sounder wave accelerated electrons circling Earth's magnetic field lines.[101] The main interest, however initially focused on the morphology of the ionosphere, which seemed to closely mirror portions of the bottom side. The "well-behaved" temperate ionosphere showed regular and slowly varying ionization. In equatorial latitudes, the ionization formed a dome over the magnetic equator, appearing to slide down the magnetic field lines so that it piled up around ±20° North and South magnetic latitude, confirming the Longitude Anomaly that had helped launch RPL's initial research late in World War II. Starting at about 45° geomagnetic latitude, the ionization transitioned to the high-latitude ionosphere, where the Van Allen belts dumped energetic particles into the upper atmosphere, causing diffuse echoes on ionograms.[102] Within the auroral zone, the satellite passed through vertical sheets of ionization several kilometers thick.[103]

By late November, the satellite had produced about 50,000 ionograms. The emphasis on morphology that had dominated the early research soon gave way to a focus on the reliability of the device and its consequences. At an aeronautical conference held at locations in Cambridge, Massachusetts and in Montreal almost exactly a year after the first results were presented, John Chapman began by explaining that in the first ten months of its operation, the satellite had received and executed more than 10,000 commands in 2,000 hours of operation of the four experiments aboard. "In this time," Chapman emphasized, "there were no equipment failures. The sole observable degradation was in reduced power output (to 63% of the output at launch) of the solar cells due largely to energetic particle damage from the artificial radiation belt produced by the Starfish test of July 9, 1962." Chapman went on to explain that the satellite had been designed "with reliability as a prime consideration," and described the results of its first months of operation. He then went on to describe the auxiliary experiments—cosmic noise receiver, VLF, and the six energetic particle counters. The last of these had been used to study artificial radiation belts created by the Starfish test and by Russian high-altitude nuclear tests in October and November 1962, one of which had produced a short-lived belt of electrons over Australia.[104]

In Warsaw the following June, Andrew Molozzi would drive home the Laboratory's approach to reliability. Ultimately, Alouette would continue functioning as designed for ten years, until it was finally turned off from the ground. Within four months of its launch, the Defence Research Board had signed another agreement with NASA for a series of follow-up satellites to provide a comprehensive synoptic study of the topside ionosphere over a complete range of solar activity.[105] In November 1965, on the verge of the launch of the second satellite in the series, John Chapman proposed to NASA to extend his group's expertise in topside sounding to the Martian atmosphere as part of the investigation into extraterrestrial life, explaining that one condition for the evolution of life on other planets was an atmosphere that shielded ultra-violet radiation. The proposed instrument was to be carried aboard the Voyager mission planned for 1967, but the Apollo I launch-pad fire in January, which killed three astronauts, delayed the mission and budget cuts in August helped cancel it.[106]

Conclusion

By the time of Alouette's launch, however, the laboratory had already begun moving away from space science to focus on the development of space technologies. The S-27 project marked a fracturing in the status of the ionogram. DRTE had used the image throughout the postwar period as a way of exploring what its researchers saw as the inherent oppositions between the natural order of high-Northern latitudes and the machinic order of high-frequency radio disruptions and communications blackouts. Alongside equipment tests and human acclimatization trials, their work had helped define the Canadian North as one of the world's hostile natures—inhospitable, threatened with invasion, and impeding the operation of humans and machines alike.

As the laboratory shifted its emphasis to ballistic missile defense in the late 1950s, its research recast the high-latitude ionosphere from a reflecting layer disrupting terrestrial communications to a hostile region threatening the operation of technologies and masking the flight of ballistic missiles. As a miniature laboratory, the satellite project was designed to investigate and document the environment of near space; as a machine that had to function there, it was shaped by emerging philosophies of hyper-reliability in extreme environments—from its design and testing, to its material details, and even the way its circuits are drawn. Not quite a mere pretext for the S-27 mission, the topside ionogram was no longer directed toward the project of reliable short-wave communications of the postwar years. Increasingly, it became a marker itself of the ability to design and build robust technologies at the height of the Cold War. Its successful production became the justification for an approach to reliability set against the powerful alternatives that the Reliability Crisis had helped create in the late 1950s.

Ultimately, the Alouette satellite and the culture of reliability it embodied did as much to undermine the status of the ionogram as to secure it. Reliability increasingly became the site around which sections of the laboratory began to position a new identity for themselves and for the nation. Here, ionograms functioned as self-referential objects, commodities divorced from the questions of distinctive natural and machinic orders that had dominated their work in the previous decade and been used to publicly justify the satellite project itself. In this way, the satellite project represented a culmination of a vision of nature and technology at the same time that it marked a shift away from the North and high-frequency communications as a focus. The satellite's enormous success would revive concerns about the ionogram in different form: How could the photographic records be produced reliably in very large quantities? How could they be analyzed on that same scale? How should they be circulated and shared? But the answers to those questions ultimately led to communications projects that would try to bypass the ionogram along with its problematic natural orders. Those efforts are discussed in the chapters that follow.

5 Infrastructures and Ionograms

Satellites reconfigured the possibilities for geophysical research in the Cold War. Ground-based observations had depended on collections of machines spread across the surface of the planet, coordinated and standardized by central laboratories who exchanged data through vast circulation networks. By contrast, a single satellite could generate global data as it orbited the spinning Earth. The ground stations and telemetry infrastructures might mirror the lattice of geophysical field stations and observatories, but the control of the data was significantly different. Whereas geophysical data had often been exchanged freely between individual research groups, providing a model for scientific collaboration, satellite data was often considered proprietary. The combination of possessiveness with the generation of huge quantities of data made it possible for non-superpower nations to reconfigure the flow of global geophysical data, but it also created enormous problems for data analysis.[1] As such, the history of these space projects points to how global data flows in the postwar period were themselves deliberately crafted and designed as both technical and political artifacts, the way space infrastructures could be open to contest and renegotiation, and the way solutions to the problems of mass production could reconfigure the natural orders they were designed to investigate.[2]

The research of the Defence Research Telecommunications Establishment had traded on arguments of a distinctive natural order for Canada—a distillation of the peculiar geophysical phenomena that extended from deep inside Earth's core to the edges of space. Outwardly, the Alouette satellite project had been presented as an effort to learn more about that natural order as a way of improving communications. Internally, DRTE had presented the work as contributing to a different machinic order of ballistic missiles and radars. Alouette would be the beginning of the most comprehensive survey ever conducted on the ionized regions of the near-Earth environment—the first project to contribute extensively to the World Data Centers established for the IGY. NASA's project manager for the mission, John E. Jackson, would consider it the most outstanding of all the Agency's international programs.[3] In the years that followed, the satellite and its successors would help shape the procedures and politics for submitting pooled satellite data, building the global infrastructures that, as Paul Edwards has noted, helped create "the world" as an object of knowledge.[4] As an attempt to reconfigure the geography of postwar ionospheric research, the project's scientific and political ambitions hinged on its ability to produce ionograms in quantities never before seen.

This chapter explores the issues involved in the reorientation of global data flows and the effects of mass production on the natural order at the core of DRTE's research. Expected to last only a year, Alouette was still producing 1,100 ionograms per day three years later—a million ionograms in total. Over its ten-year lifespan, it would produce two million records.[5] Through the mass-production system designed to ensure its integrity, the ionogram became the proving ground for techniques of broad interest to the Cold War sciences and engineering, particularly around questions of automation and real-time processing. But automation, when applied to analysis *en masse* of the ionogram, effaced the very complexity that researchers had argued characterized the ionosphere above Canada and tied the graphs to uniquely violent communications disruptions. In addition to shaping the production and analysis of ionospheric data, those techniques informed projects within DRTE that ultimately sought to circumvent the ionogram altogether, automatically exploiting meteor trails for communications or using real-time ionospheric sounding to select usable radio channels. They formed an alternative, supra-national vision of the natural and the machinic orders—machines capable of instantly and automatically adapting to the conditions of the upper atmosphere, functioning in unison with it while avoiding its representation, cutting out altogether the mediating role of the ionogram once and for all.

Those developments evolve over the chapter's three main sections. The first section deals with debates about how the ground station network should be organized, who should control the satellite, and how data from the satellite should be shared. The second explores how these considerations formed part of the technical design of the satellite, specifically requiring a system for mass producing ionograms in Ottawa from data gathered around the world. The third focuses on the problems of data analysis that this system produced and the new reading techniques that were devised to analyze the overwhelming number of ionograms Alouette produced. It examines how projects to automate data analysis effectively erased the very ionospheric phenomena that previous regimes had struggled to capture and that DRTE had made central to its work. These same techniques would later be used to develop alternative communication systems—parallel machinic orders—that cut the intricate links between hostile nature and failing machines. In doing so, the chapter retraces some of the questions about production and interpretation that were investigated in chapters 2 and 3, but with an eye toward how these procedures cut orthogonally, eliminating or obscuring the natural and machinic orders that featured in that earlier story. That discussion prepares the groundwork for the final two chapters, which examine how those associations between nature and machines were eventually overturned, first by a set of technologies that circumvented the natural or exploited its anomalies, redefining its hostility, and second through the development of communication satellites.

Networks of Negotiation

As Walter McDougall noted in his classic study of the American space program, space posed two of the overarching international problems of the twentieth century—how to control arms development within a context of weapons escalation and mutual distrust; and how to manage non-territorial regions that included Antarctica, the oceans, and the atmosphere.[6] In addition to those pressing issues, space activities raised a related problem, resolved incompletely during the IGY, of how data should be produced and shared when the nature of space missions and the structures of space activity spanned nations and continents. Although scholars have focused extensively on spacecraft and launch devices as the products of international collaborations, satellites themselves were only the most sensational elements in an enormous and largely invisible infrastructure built to locate them after launch, track them during orbit, and receive their data from space.[7]

Although several methods were developed for those tasks, space infrastructure came to rely principally on radio-frequency signals that linked the satellite to a global network of ground stations. In using frequencies that cut through the ionosphere and atmospheric noise, the ground stations were forced to rely on line-of-sight transmissions. Since even relatively flat portions of the planet only reveal about 15° of a given satellite orbit in any direction, the ground stations formed a vast chain, with each station successively acquiring and tracking the satellite as it rose above horizon, and then "handing off" to the next station as it disappeared from view.[8] The design of the satellite, particularly its ability to store information, was related to the size of the network needed to support it. Satellites that lacked the ability to store data onboard required more stations on an orbital pass to capture their data as it was beamed back in real time. An equatorial orbit would require a single ring of stations; as the orbit's inclination (its angular separation from the plane of the equator) grew, that ring broadened into a thicker and thicker swath of the earth as it rotated beneath the orbiting spacecraft.

Although individual orbiting satellites could generate data from around the globe, their ability to produce *records* depended on the network of ground stations that received the raw data and recorded it on magnetic tape for future processing. The technical details of both objects, like the politics of their creation and collaboration, were deeply interrelated.[9] The design of the satellite, its mission, and the choice of its orbit were of a piece with decisions about diplomatic overtures, geopolitical rivalry, and strategic alliances.[10]

The technical features of the Alouette satellite significantly shaped the infrastructure of the tracking and data-acquisition stations. To reduce complexity, Alouette's engineers had decided to avoid storing data on the satellite itself. One consequence of that decision was the need for an extensive global network of telemetry and receiving stations. The inclination of Alouette's orbit also meant that the ground network would have to cover almost all latitudes. According to the formal agreement, NASA would provide part of that network through its existing tracking and telemetry system.[11] By September 1962, when Alouette was launched, NASA had more than thirty ground stations—each made up of three large antennas and three trailers housing telemetry and communications equipment—spread across five continents and numerous islands and connected by about two million circuit miles of cable on land and on the ocean floors.[12]

The network had been started before NASA was created, in anticipation of an American satellite program. Seven stations were built along the 75th Meridian from Maryland to Chile, and there were additional stations at San Diego, near Johannesburg, and at Woomera. The system's designers felt that this would give a 90 percent chance of capturing every pass of a satellite at 500 kilometers or higher. The layout mirrored the initial dreams of ionospheric physicists who hoped to create a string of field stations along a single meridian of longitude, allowing researchers to extrapolate results for locations anywhere in the world. After some modification, the original ground stations were used to track Sputnik, and were transferred completely to NASA after it was officially formed in July 1958.[13]

For Alouette, the existing system would have to be expanded to accommodate the higher-inclination orbit. Canada would build four of the new stations, including a primary tracking and data processing facility at DRTE. Those stations would provide radio-frequency links for Alouette's housekeeping information, as well as the ability to abort the rocket after launch, command the satellite, and hand off the spacecraft to maintain almost continuous contact during its orbit. To locate the satellite, the system used a radio interferometry system based on ballistic missile tracking developed by the Naval Research Laboratory at White Sands.[14] Once the satellite was located, the stations measured its angular position and velocity and used real-time calculations to alert non-acquiring stations about when and where they could expect to find it. But only the Ottawa station could produce ionograms from the telemetry tapes, and only that station had complete control over the spacecraft. To avoid the need to hand over sensitive command codes to other countries, and also to conserve its batteries, the satellite controller in Ottawa would program the satellite to turn on as it flew over few specific sites.[15] Together, both the satellite—as an object that could generate raw data around the world—and the details of the data handling reconfigured the production and flow of ionograms that had characterized ionospheric research since World War II.

The creation of that global network was itself complicated and shot through with political concerns much larger than those surrounding the design of the spacecraft. In early 1961, for example, NASA officials requested that all references to the Falkland Islands telemetry station, which was run by the British as part of the Alouette program, be deleted from public documents—ostensibly to avoid concerns that the United States might take sides in the dispute between Argentina and Britain concerning those islands. American ground station equipment had to be shipped first to the United Kingdom, loaded onto British ships, and then transported to the Falklands as British equipment. The thorniest negotiations, however, concerned Soviet participation in telemetry and data acquisition.

As the Alouette ground network began taking shape in 1960 and 1961, officials at both NASA and DRTE began thinking seriously about Soviet involvement in the project. Under Eisenhower, the US Department of State had favored direct cooperation with the Soviet Union in space research. NASA's Director of International Affairs, Arnold Frutkin, was open to proposals for scientific collaboration on space projects, but presented them to the Kennedy administration as part of a larger plan for using spurned American overtures to diplomatic advantage.[16] In the case of Alouette, there were technical advantages: a ground station in northern Russia, for example, would increase coverage of the polar ionosphere. It would also add to synoptic investigations by contributing to a nearly uninterrupted line of ionospheric and satellite stations stretching from just north of Antarctica, over the Americas along the 75th meridian, across the North Pole, over the Soviet Union, and nearly to Manchuria. The fact that the satellite was Canadian was a diplomatic advantage. From the standpoint of American foreign policy, the Canadian satellite and its data, already a much-trumpeted symbol of international cooperation, would provide shielded points of contact between the superpowers, avoiding the appearance of outright overtures to the Soviet Union while signaling the increased cooperation that John F. Kennedy had called for in his inaugural address in January 1961.

Canadian officials, for their part, recognized the mediating role they might play in a possible rapprochement—something they had already attempted in late 1958 and early 1959, when the UN was locked in a complex struggle over international cooperation and the peaceful uses of outer space. The US, seeking to cloak its military interests by leading the effort for peaceful cooperation, had proposed an ad hoc committee for the purpose—the future UN Committee on the Peaceful Uses of Outer Space (COPUOS).[17] The Soviet Union saw the move as a ploy to limit its own involvement in space research and refused to participate because of the committee's composition.[18] The United States, supported by the United Kingdom, was adamant that Soviet obstructionism and suspicion should not prevent the committee from doing its work and pushed for it to convene as soon as possible, with or without Soviet participation. Poland and Czechoslovakia had sided with the Soviet Union. The Canadian position, spelled out by Norman Robertson of Canada's Department of External Affairs, was that the Soviet Union was the leading nation in space activities and that its participation might be essential to international cooperation in space—a view supported by India and the United Arab Republic (a short-lived political union between Egypt and Syria, cobbled together to thwart a communist takeover). Robertson saw the pressure by the United Kingdom and the United States as a propaganda exercise against the Soviet Union and an attempt to gain a narrow military advantage by excluding the Soviet Union from this facet of international collaboration.[19] Fresh from its triumph in helping to defuse the Suez crisis, the Department of External Affairs had an interest in acting as a broker for these tensions while exploring how to serve Canada's future interest in space activities. Favoring a dynamic American space program and sympathetic to the American position, but wary of a military competition over space that would alienate the Soviets and jeopardizing future participation, Robertson looked for a way to initiate the work of the committee without causing the Soviets to withdraw entirely and inflame tensions.[20]

Robertson had met with members of the Defence Research Board and the National Research Council on January 21, 1959 in the East Block of the parliament buildings to discuss the question. The meeting quickly shifted to discuss Canada's priorities for future space research. In the summer of 1958, one of Robertson's undersecretaries, Douglas LePan, had developed an allied proposal designed to address concerns that Canada, if left out of space development, would find its own interests threatened or restricted. He formulated a proposal for Robertson and Prime Minister John Diefenbaker that would ensure third-country claims over outer space by placing space activities under international control, while promoting Canada as a potential site for international launch facilities at Churchill—in 1958 the only potential site for the West to launch satellites into polar orbit.[21] Representatives of both the Defence Research Board and the National Research Council favored collaboration through existing bodies rather than through the new UN committee. In particular, they looked to the Committee on Space Research (COSPAR)—which had been set up in November 1958 under the International Council of Scientific Unions, the body responsible for organizing the IGY—to continue space-research programs.[22] The fear was that the work of the COSPAR Committee, which had been formed "in a purely scientific atmosphere without political considerations" and which already represented the world powers with interests in space research, would be taken over and impaired by "political wrangling over co-operation in the United Nations." Robertson had proposed that space science, rather than launch technologies, would give Canada and other small nations an opportunity to push for international control of outer space while collaborating with the United States to gain preferential and occasionally exclusive access to advanced technology.[23]

The consensus that emerged from the January meeting, communicated in a confidential memo sent from the Department of External Affairs to UN officials, to embassies in Washington, London, Paris, and to NATO headquarters, expressed reluctance to see the UN committee meet. It asked Canada's UN representative to "take the position in your discussions with the United States and the United Kingdom delegations that we are opposed to convening the committee merely as a gesture to show the Russians we mean business and are not intimidated by their boycott." Instead, in the event the UN committee was convened, preferably after the next COSPAR meeting in March, the Department of External Affairs provided a series of working recommendations to avoid either antagonizing or alienating the

Soviets: that the committee should seek consensus rather than voting on certain issues, that reports should avoid isolating the Soviets by placing their position in minority views, and that there should be no railroading of its final report through strength of numbers.[24]

The negotiations over the Alouette ground stations in the spring of 1961 revived the possibility of fuller cooperation. In March, members of DRTE began working out the details of possible Soviet involvement.[25] NASA, cautious to avoid direct negotiations with the Soviets and citing the Canadian origins of the satellite, urged Canadian officials to make contact. During the late 1950s and the early 1960s, this intermediary role for Canadian officials, sufficiently trustworthy for the Americans and sufficiently inoffensive to the Soviets, was often invoked, as their role in the Cuban missile crisis demonstrated.[26] Assuring their superiors in the Department of National Defence that they would "carefully avoid anything which may be construed as a Canadian overture to Russia," DRTE officials suggested that John Chapman, the director of the Alouette Program, might approach Soviet delegates at the 1961 COSPAR meeting in Florence in early April.[27]

Chapman had risen quickly through the ranks of DRTE, from summer student in 1948 through increasingly senior positions in the Radio Physics and Communication laboratories to the position of Deputy Chief Superintendent of the Establishment. His proposal to the Soviets would be for a telemetry station located at Tiksi Bay and outfitted with its own data-reduction equipment, capable of producing ionograms locally. To avoid the thorny question of "interrogation of the satellite" (a telling metaphor) by the Soviet Union, DRTE officials proposed to turn the satellite on during predetermined passes over the North Pole. The data would then be exchanged in the form of topside ionograms (rather than magnetic tapes), following the precedent of the IGY. DRTE would even consider curtailing soundings elsewhere on the globe in order to conserve satellite power if it meant accommodating Soviet participation.[28]

The entire plan unraveled in two days in early April, before it ever reached the Soviet delegation. The causes were a mix of technical and political calculation. There was firstly the problem that the Soviets might not have data-reduction facilities available for a Northern station, in which case they would have to send their data to DRTE for processing. Those data had to conform to the proper format (Ampex FR 100 standard), forcing the Soviet Union into the politically unpalatable move of adopting American tracking and telemetry equipment due to time constraints. The need for a communications link between Ottawa and the Soviet station also presented language difficulties. Finally, handing over control of the satellite to the

Soviets meant releasing satellite command frequencies and command codes that could "compromise" American satellites or space probes.[29] In the end, Chapman and his colleagues advised NASA that the situation needed further study.

In parallel with the development of the tracking network, Chapman and his group had started thinking early on about the possibilities topside ionograms provided for international exchange. Ultimately, they would explicitly model their system on nuclear research.[30] Chapman, who as supervisor of the Alouette project handled requests for data acquisition, wrote to his NASA counterpart in October 1964 explaining how "the problem of releasing data from long-lived satellites is beginning to come to our attention from other directions."[31] At various international conferences throughout the preceding year, scientists from India, France, Japan, Russia, and Germany had all approached Chapman about acquiring Alouette data. In turning them down, Chapman initially invoked arguments about limited power of the satellite. "I can't really believe that argument myself now," he confided in late 1964. "Those of us in geophysics," he explained, "are accustomed to thinking in terms of the exchange of data à la IGY," in other words, with unlimited restrictions. Nuclear physicists, "with their equally costly machines," he added on the other hand, "have no such tradition."[32]

The protocols of nuclear science would guide members of DRTE and NASA from late 1962 through 1964 as they worked out the basic procedures for acquiring and processing topside data. Participants now were divided into principal experimenters (DRTE and NASA) and everyone else. Only the principal experimenters would be able to publish topside ionograms or "routine tabulations" during the first year after data acquisition, although other agencies (among them the British Radio Research Station at Slough) might be granted these privileges on the basis of special relations and significant contributions. It furthermore became standard practice to ensure that ionograms met a "minimum standard" before they could be published or used by the agency that had produced them.[33] In addition to the imposition of a standard format, DRTE had the task of examining and commenting on the quality of sample ionograms from each data-processing center. By 1968, tempered by practical considerations and political circumstance, the system was a mix of inclusiveness and restriction: twenty-one ground stations under the control of Canada, the United States, Britain, Japan, France, and Norway. With the exception of large portions of the Soviet Union and Asia, this gave coverage of most of the world's major land masses. (See figure 5.1.) It also formed a colossal catch basin of global ionospheric information with Ottawa at its center. The wealth of the resulting records would

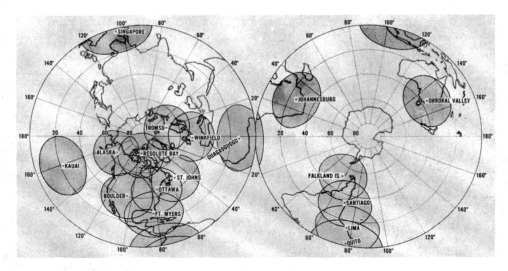

Figure 5.1
Station coverage for the Alouette I and II ground stations (shown in polar projections). Most of the world's landmasses were covered. Notable exceptions were the Soviet Union and much of Asia. © 2016 IEEE. Reprinted, with permission, from C. A. Franklin, R. J. Bibby, and N. S. Hitchcock, "A Data Acquisition and Processing System for Mass Producing Topside Ionograms," *Proceedings of the IEEE* 57 (1969).

transform the laboratory, soon pushing it away from the narrow but pressing concerns of the polar ionosphere and into the broader field of synoptic studies. For this to happen, though, the laboratory needed a mode of ionographic production that matched the scale of the new station network.

Mass Production and the Ionospheric Record

The political maneuvers and technical stipulations for the satellite ground network underscored the value of having a large, unbroken stream of ionographic records flowing from the various stations. At stake was what we might call—following the tradition of geology—"the ionospheric record."[34] Like geology, that record made ionospheric research a science of the archive.[35] The history of ionospheric research at DRTE had in many ways been an attempt to establish the integrity of a part of this archive against the vagaries of equipment, environmental conditions, and human operators. The adoption of the standard ionosonde, closer supervision of technicians, and new interpretive practices for the ionogram all attempted to secure the steady flow of records and their stable movement between field station and central laboratory. In the early 1960s, the physical ionograms, the actual strips of photographic film, were themselves deeply implicated in these concerns. The practice of forwarding original photographic records to World Data Centres for duplication, for example, was fraught with the anxiety of loss or destruction, not to mention the stigma of dependency. In August 1962, writing to his superiors, Frank Davies effused that DRTE finally possessed the facilities to duplicate film records in house, which made it "unnecessary for the original ionograms to leave the establishment, and avoiding the loss of original data in transit."[36] The records were now stored in a growing "library of ionograms," a comprehensive archive of records documenting the state and evolution of the world's ionosphere.

In ensuring the integrity of this catalog and Alouette's contributions to it, researchers and officials at DRTE looked to the resources and rhetoric of industrial production. From the earliest stage of the satellite project, the satellite's Chief Electrical Engineer, Colin Franklin, had tied the value of Alouette to its ability to mass produce records.[37] One of the criteria that researchers identified for its enormous success was its ability to make iono-grams rapidly available to international researchers.[38] In the 1960s, data from scientific satellites were considered the property of the principal investigators, usually for a period of one year, after which the data were made available to the wider scientific community.[39] In exchange for operating telemetry and tracking stations for Alouette, participants were effectively given the status of principal investigators, requiring that the magnetic tapes be processed quickly and at scale, with the resulting ionograms distributed to the col-laborators. Franklin and his team had designed those goals into the satellite itself, equipping it with a high-power transmitter to overcome the ambient noise and spatial losses in the upper atmosphere, and to produce a teleme-try signal that could be received, demodulated, and displayed by relatively inexpensive equipment.[40] If the Alouette project was to be truly global, Franklin reasoned, with ground stations from Singapore to the Falkland Islands, the satellite would have to be incorporated into a system of indus-trial production in which ionographic data traced a stable and uninter-rupted path from spacecraft to magnetic tape to readable ionogram.[41]

The large-scale production of ionograms depended on strict labor prac-tices enforced at ground stations throughout the sounding network. In commenting on the loss of some satellite data from the Resolute station, high in the Canadian arctic, one official explained: "Experience at DRTE has shown that tight quality control is necessary if outages [of data] are to be kept to a minimum. We have been successful in this only when quality control and technical direction have been consolidated." In early 1963, Franklin and his colleagues had devised a system at the Ottawa ground station that they believed epitomized this union of quality control and technical direction. (See figure 5.2.) The system claimed to treat the processing of ionograms as "a unified problem involving the topside sounder, the spacecraft telemetry and command links, the ground stations, and the data processing center."[42] The upper panel in the figure shows schematically the path of ionographic data in passing from tracking antenna to magnetic tape. Apart from addressing a number of technical difficulties, the system was significant for the way it introduced numerous "monitor points" throughout the data-recording system to ensure the quality of the final tapes. The tapes were then fed into one of two processing systems (shown in the lower panel), which generated the photographic records. In this way, by concentrating the production of film ionograms at Ottawa and enforcing quality-control measures throughout the ground network, DRTE hoped to reduce the haunting problem of data variation from the outlying stations.

The industrial inspiration also embodied concerns about its human users. Throughout their exposition of the system, the system's engineers openly implicated humans in the "unified problem" of ionographic processing. "In designing an optimum system for displaying ionograms," Colin Franklin explained, "there are a number of relatively straightforward electrical, mechanical, and optical problems to be solved. However, there are also some important questions to be answered concerning the display and visual detection of ionospheric echo pulses in noise. Many of these questions are interrelated and some involve subjective and physiological factors which are difficult to define."[43]

For an electrical engineering team, full-blown psychology may have been beyond the pale. But an operationalized physiology was not. Take the speed of the advancing film: since the width of the ionogram was fixed by the distance between sprocket holes on the film (about 20 millimeters on 35-mm film), the film velocity determined the aspect ratio of the ionogram— the ratio of its length to its width. In setting this velocity, the designers considered what they called an "experimenter's requirement." "The length of the ionogram must be at least twice its width," they posited. "This is because aspect ratios of <2:1 introduce basic difficulties in visual scaling of ionograms." In other areas, something very close to a stripped down psychology seemed to take a hand. In discussing signal-to-noise ratios the designers cited a series of studies conducted using Alouette II ionograms. These suggested that the point at which human observers could pick ionospheric traces out of the noise (called the "detection threshold") was

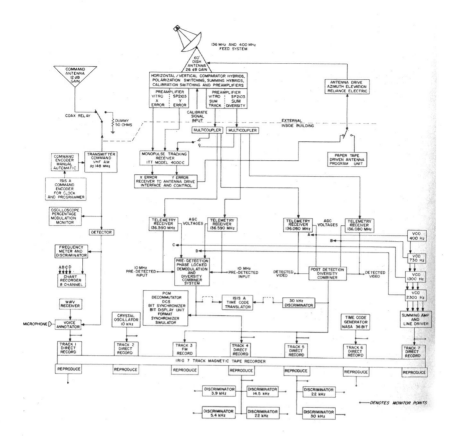

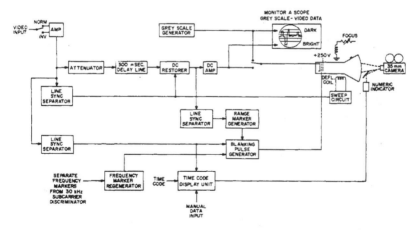

Figure 5.2

Franklin's mass-production system for ionograms. The upper diagram depicts the data-processing system that transformed telemetry data into high-quality magnetic tape recordings. The lower diagram depicts how those recordings were converted into photographic ionograms. © 2016 IEEE. Reprinted, with permission, from C. A. Franklin, R. J. Bibby, and N. S. Hitchcock, "A Data Acquisition and Processing System for Mass Producing Topside Ionograms," *Proceedings of the IEEE* 57 (1969).

considerably lower for ionograms than for A-scan records (discussed in chapter 2). "It seems," the designers ventured, "that the eye correlates the echo pulses appearing on side-by-side scans and searches out areas of the record where lines of echo points occur, and disregards the random noise points."[44]

Despite these considerations, the system would come under indirect attack for its neglect of the problem of data reduction. In its aim of mass producing high-quality ionograms, Franklin's system was wildly successful—perhaps too successful. With two processors running, it generated about 200 ionograms per hour. By the fall of 1968, after roughly six years of operation, the system had produced approximately 2.3 million ionograms.[45] The processing requirements alone taxed the human resources of the establishment. "Allowing for reruns, special processing, quality control checks, etc.," the designers explained, "it is necessary to operate the data center 70 hours per week to process 6 hours per day of data transmissions from these spacecraft." The scheduled launch of three more ionospheric satellites in the coming years and the proposed international adoption of Franklin's system as part of collaborative analysis with Britain, Australia, France, India, Japan, New Zealand, Norway, and eventually Finland demanded changes. "Higher tape playback speeds and increased automation are therefore required."[46] In the meantime, though, the preservation and expansion of the ionospheric record would require enforcement of strict labor relations and practices among the data analysts now awash in topside ionograms.

Mass Interpretation

Initially, the analysis of the topside ionograms focused on unusual or distinctive phenomena not present on ground-based records: plasma and cyclotron spikes and unexpected sheets or columns of ionization around the auroral zone.[47] But it soon transformed the work of DRTE in several ways, the most important of which was automation. The need to process masses of ionograms quickly would ultimately erase the distinctive natural phenomena that DRTE had made the focus of ionographic interpretation for a decade, just as parallel communication projects would try to devise new machinic orders to bypass it altogether.

The impulse toward automation initially came out of the need to calculate electron density ($N(h)$) profiles, one of the thorniest and most pondered problems in ionospheric research during the early 1960s. Ever since the late 1920s, these profiles had borne an important relationship to ionospheric theory.[48] The interpretation work owed much to Edward Appleton, one of the earliest users of the ionosonde, one of the principal authors of magneto-ionic theory (the basis of ionospheric propagation theory and analysis techniques), and the creator of one of the most enduring models of ionospheric layer formation and dynamics. Appleton's efforts would earn him a knighthood in 1941 and a Nobel Prize in 1947. His student and colleague J. A. Ratcliffe, who shared with Appleton an appreciation for the ionogram's richness and versatility, would carry forward the work on density profiles. During a sabbatical in Washington in 1952, Ratcliffe developed a method of using ionographic data to estimate the "integrated electron content" of the ionosphere up to the peak of the F2 layer. After returning to England, he set up a major computing effort to derive $N(h)$ profiles, first using table-top machines and then employing EDSAC, the digital computer built in 1949 by the Cambridge mathematician and wartime code breaker Maurice Wilkes.[49] Ratcliffe's efforts, along with those of his colleagues and students, would help turn postwar Cambridge into a leading center for the interpretation and manipulation of ionograms.

Leroy Nelms was an obvious choice for the job of calculating $N(h)$ profiles from the Alouette records. A recent recipient of a physics doctorate from Cambridge, Nelms had learned his trade under Kenneth Budden, one of the early developers (alongside Ratcliffe) of real-height analysis. At Cambridge, Nelms had acquired an abiding appreciation for the complexity and promise of the ionogram. Interpreting these records was, for him, both an art and a science. The science involved detailed knowledge of theories of radio propagation, instrument characteristics, and models of ionospheric structure, all of which workers needed to command when examining the records. But in addition, an ionospheric physicist or a communications engineer also had to deploy more creative capacities. Nelms held that the "art" of ionographic interpretation involved using the human capability for reasoning and speculation to consider various propagation scenarios and to unravel the physical processes behind the ionographic traces. Interpreting the ionogram, Nelms would later observe, involved "using the human mind in a very exotic way."[50] In late 1962, as Nelms began his work on density profiles, he would use these imaginative demands to distinguish the practice of ionospheric physics from the routine tasks of data reduction and data processing.

Building on his training at Cambridge, Nelms began compiling a new form of density profile. Previously, $N(h)$ profiles had been highly generic representations, directed toward general theories of ionospheric formation and rarely bearing markers of specific location or time. Occasionally they were complemented by more localized maps showing changes in electron density or critical frequencies, and superimposed on geopolitical entities such as Canada or the countries of the Northern Hemisphere. Because the Alouette satellite traveled nearly from pole to pole, it became possible to instead represent the ionosphere in broad cross-sectional swaths. Nelms' charts took the form of synoptic profiles—contour maps showing the variation of electron density with latitude. (See figure 5.3.) The records stood in stark contrast to Jack Meek's careful correlations in 1954 of highly localized geophysical phenomena. In the new presentation, nations, boundaries, bodies of water, geophysical characteristics, and the territory itself were largely absent. The graphs suggested that the border between the mid-latitude and high-latitude ionospheres was located at about 70° geomagnetic latitude. Members of DRTE would still read geography into these graphs, explaining how "abruptly, at about Toronto, the smooth mirror [of the ionosphere] becomes rough."[51] More strikingly, the discussion emphasized new "near-permanent" structures of the polar ionosphere, particularly a series of peaks and troughs that were accentuated during ionospheric storms.[52] But apart from precise geographic coordinates at the endpoints, the graphs themselves were stripped of any explicit connections to Canadian territory.

This separation itself mirrored a transformation in the practice of ionographic analysis and in the figure of the data processor. From the strict standpoint of routine data processing, the ionogram had always contained too much information. Cutting through this complexity was the primary function of the scaling rules elaborated by Jack Meek and Frank Davies in

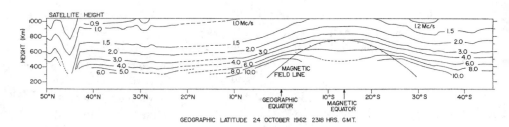

GEOGRAPHIC LATITUDE 24 OCTOBER 1962 2318 HRS. G.M.T.

Figure 5.3

A contour map generated from Alouette data. Using isolines of critical frequency, the charts show the variations in frequencies (and therefore electron density) as a function of both height and latitude. Note the abrupt change in the ionosphere at about 45° north latitude as the satellite enters the "polar ionosphere." Source: Leroy Nelms, "Ionospheric Results from the Topside Sounder Satellite Alouette," in *Space Research IV* (Proceedings of the Fourth International Space Symposium), ed. P. Muller (North-Holland, 1964). Copyright Elsevier.

the early 1950s, rules directed at isolated field operators and discussed in detail in chapter 3. But the work at the field stations had involved carefully measuring and recording adjunct data, such as radio reception conditions, auroral observations, and magnetic readings, that were then combined with ionospheric data to provide a relation between atmospheric phenomena and radio failures. Despite the professional, social, and epistemic divisions between laboratory and field, the men operating the field stations of ground-based research had been responsible for a wide range of competences stretching from clerical diligence to instrument operation and scientific observation. In this, their work had embodied the mediating role of ionospheric research as a discipline straddling communications and geophysics.

The Alouette project transformed this situation in at least three ways. First, it focused narrowly on the topside ionosphere, eliminating the simultaneous recording of radio conditions and auroral displays which were the province of bottom-side investigations and which had helped match natural order with the behavior of machines. Second, it moved control of the instrument and the act of experimentation out of the hands of observers and data analysts working at scattered outposts and into the hands of engineers and satellite controllers in Ottawa. Finally, it translated the dispersed practice of data reduction from the unruly and heterogeneous field stations into the homogeneous and controlled environment of the laboratory. In doing so it redrew labor relations along the (often) coextensive lines of gender and professional status. In separating experimental control (under engineers) and analytic control (under human data processors), the Alouette system effectively divided a set of tasks that had previously belonged to the station operator, and simultaneously placed routine data analysis in the hands of women.[53]

The group in charge of providing Alouette data for the density profiles was the Synoptic Studies Section of the Radio Physics Laboratory. (See figure 5.4.) The Synoptic Studies Section performed two important functions: it comprehensively analyzed data already collected (determining the spatial and temporal extent of ionospheric absorption, for example), and it conducted routine data scaling and processing for researchers in other sections. The bulk of the section's personnel were involved in the second of these. Apart from its supervisor and one other member, the Synoptic Studies Section was entirely composed of women, referred to (in a term that cut across the wider sweep of national and disciplinary boundaries) as "girls." Although women held a number of positions within DRTE (the accomplished spectroscopist Luise Herzberg was a member of the Upper Atmospheric Physics Section), data analysis at DRTE mirrored the gendered practices of

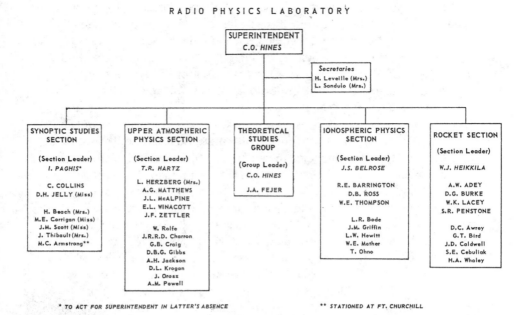

Figure 5.4

An organizational chart for RPL showing the Synoptic Studies Section (extreme left). Honorifics were attached only to women (Miss or Mrs.). The section was disbanded shortly after the launch of Alouette I, but its members continued to perform the same data-reduction duties in their new assignments. Source: Defence Research Board, "DRTE Scientific and Administrative Organization," DRTE Publication No. 1037, January 1960. LAC, Record Group 24, volume 9, file DRBS 100–22/0, part 2. © Government of Canada. Reproduced with the permission of Library and Archives Canada (2016).

data analysis elsewhere in the physical sciences. As in many other areas of postwar physics and electronics, this careful work was coded female, instituting a largely gendered division of labor within laboratory and discipline alike.[54]

Because data-reduction facilities were concentrated in Ottawa, and because routine data analysis was carried out almost exclusively by female staff, the Alouette project effectively created a female laboratorial counterpart to the male field-station operator. Data reduction began with the projection of ionograms (conveniently stored on 100-foot rolls of 35-mm film) onto viewing tables. Here, under the flood of records from Franklin's mass-production system, the graphs were interpreted according to the ion-ographic taxonomies detailed in chapter 3 and "scaled" according to the appropriate rules set out in the instruction manuals in order to render two measurements—the virtual height of the ionospheric layers and the corresponding critical frequency of penetrating radiation—along with additional information about time and location. At this point, the emerging computational infrastructure of postwar research came into play. IBM card punches, verifiers, and sorters were used to record the requisite data on punched cards. Initially these were fed into electro-mechanical calculators (which often jammed in hot conditions) and later into computers to generate the $N(h)$ profiles that fed the theoretical investigations of upper atmospheric physics.

From 1960 on, the Communications Laboratory made increasingly comprehensive attempts to automate data analysis across a wide range of fields.[55] Initiated in October 1960, it aimed to use the emerging material infrastructure of digital computers to help in the processing and reduction of masses of analog data—varying voltages, frequencies, and currents from radar echoes, communications traffic, and the anticipated calculation of electron density distributions from the satellite project.[56] After the launch of Alouette, the efforts of the project came to concentrate on the steady build-up of unprocessed telemetry data. By April 1963, approximately six months after the launch of Alouette I, only about half of the 2, 858 recorded telemetry tapes (about 1,400 miles of taped information) had been processed.[57] Researchers focused instead on the more basic task of extracting automatically "those data which are pertinent to the experimenter's interest" and manipulating these using digital computers.[58]

By 1964, the satellite project had become the proving ground for wider Defence Research Board efforts in automation. The objectives of the project were to develop automated methods of extracting usable data from large amounts of recorded data and to find ways to transform subsets of those data into forms that could be analyzed by computers. Since ionograms contained a wide array of electromagnetic phenomena, only a small fraction of which was useful for a given application, the satellite project was seen as a test case for automated data selection (mainly through pattern recognition) in analogous applications, including radar tracking and target discrimination.[59]

In June 1969, after a succession of projects aimed at the problem of automated data analysis, G. E. K. Lockwood proposed pushing automation to new levels. A radio physicist in RPL, Lockwood had started out the decade in the applied propagation section of the Communications Laboratory, where he specialized in ultra-high frequency (UHF) radio propagation for long-range detection and tracking of satellites and ICBMs. With the launch of Alouette I, he had lent his interests to ionospheric propagation, analyzing the plasma and cyclotron spikes that characterized topside records. After launch of the second Alouette satellite, however, he had joined Leroy Nelms in the Upper Ionosphere Group of RPL, dedicating himself more fully to the generation and analysis of the new contour maps that were quickly becoming the representation of choice at DRTE. It was in the generation of these maps, in the nuances and uncertainties of process, that Lockwood saw a useful role for automation.

Like so many of his colleagues in the Communications Laboratory, Lockwood sought out automation partly for the increased productivity and data use it promised. He had seen its results at work in the field of radar tracking, for example, where Communications Laboratory scientists had designed a system to record data ready-made for statistical analysis by a general-purpose computer.[60] L. E. Petrie, with whom Lockwood had collaborated on Alouette ionization studies, had begun using an IBM 650 to calculate frequency predictions over a large number of North Temperate and Arctic propagation paths on varying frequencies.[61] But Lockwood was equally, if not more, motivated by what he saw as the fundamental shortcoming of existing reduction systems: their tolerance for human error.

Error per se was not foreign to the practice of ionospheric reduction. The elaboration of rules of ionospheric analysis and the introduction of the automatic ionosonde were in large part designed to deal with the interpretive errors of isolated technicians working in scattered ionospheric field stations under trying conditions. Even within the centralized reduction system of the Alouette project, error was an accepted, if lamented, element of the daily business of ionospheric research. It was not human error itself, then, but rather the absence of any check on interpretive mistakes that preoccupied Lockwood. The problem was well represented in the current reduction system, but it signaled a more general problem in which the phenomena on scientific records were underdetermined by the physical conditions, making multiple interpretations possible. Checking the consistency between the various interpretations and the final result was too complex a task for human inspectors.[62]

For Lockwood, this general condition was clearly represented in the production of $N(h)$ profiles, in which he had been immersed for two years. Alouette's high-latitude ionograms presented the problem of multiple interpretations in particularly acute form. The prevalence of spread echoes on many of Alouette's records and the technical and conceptual difficulties in ionographic production and interpretation meant that, "portions of the data are often fragmentary, that is, incomplete over just the range that would allow a more direct [i.e. unique] interpretation."[63] Human data processors were therefore often forced to furnish insufficiently constrained interpretations and measurements without a check for consistency between their work and the final result—the $N(h)$ profiles generated later and often in different locations using a computer. The problem, then, was not so much that data reduction took too much time. Rather, in a situation mirrored in other areas of physical research, the labor-intensive work of analyzing data had created a system in which the ionogram was decontextualized, separated—physically, temporally, even socially—from the electron density profiles it helped produce.[64]

The situation, Lockwood continued, was more dire still. Even if an ionogram and an $N(h)$ curve could be placed side by side at the time of scaling, Lockwood reasoned, "there is no convenient way to relate the shape of the profile obtained to the specific interpretation of the ionogram." Lockwood concluded that "a data-reduction system without this limitation is highly desirable because of the large number of topside ionograms whose scaling and interpretation may unknowingly be erroneous."[65]

The root of the problem, for Lockwood, lay in human limitations. The absence of any comparison, in the interests of consistency, between these two graphic forms—the ionogram and the $N(h)$ profile—represented for Lockwood a fatal flaw in the present system, a door opened wide to unchecked proliferation of errors through the corpus of ionospheric data. He therefore proposed a new system that would eliminate the failings of the current system and, with them, the shortcomings of its human operators.

Going Digital: Reducing Ionograms to Cybernetics

Within the Communications Laboratory of the 1960s, where Lockwood had worked, error had been a topic of intense study. In the mid-1950s, concerns about it had dominated considerations of the JANET system, an innovative radio propagation system (to be discussed in the next chapter) that used meteor trails as reflection surfaces. Through their work on communications systems and radar tracking, groups within the Communications Laboratory had begun linking the idea of communications itself with the integrity and faithfulness of information and even the efficiency of "man-machine interactions."[66] Papers on information theory, verbal communication and sensory perception pursued these concepts in more abstract form, even examining the conditions for the possibility of communications and for the reliable transmission of data.[67] Studies were generally directed toward the limits of technology and the unpredictability of natural phenomena rather than the shortcomings of humans. They nevertheless involved two techniques that would be central to Lockwood's thought later in the decade. The first was the use of comparison, or feedback, in the elimination of error. In its research on encoding binary messages, for example, the laboratory's Communications Research Section had developed what they considered "an analogue of naval signal flags" that was enhanced by "the added feature of redundant coding to detect and correct errors."[68] Similar work on satellite command and telemetry added notions of "adaptive variation"—an elaboration of contemporary concepts such as "decision feedback" or "variable redundancy encoding"—in which the system would evaluate and adapt to its own error rates, varying the level of redundancy or information transfer accordingly in order to eliminate errors altogether.[69] The technique involved digitization and was carried out primarily so that comparisons and corrections could be made quickly, in real time, using computers.

Together, the two fields suggested feedback mechanisms as the safeguard of accuracy. They also suggested that digitization and computers might be called in where human abilities fell short. Although Lockwood's interests were with humans rather than machines, he looked to feedback and machine-aided correction for insight. Drawing on earlier digitization work, Lockwood turned to the world of digital information and error correction to envision a melding of human and machine.[70] In order to analyze consistently the often incomplete and fragmentary data of ionospheric physics, he asserted that "a system is required that combines the abilities of the human operator for pattern recognition and decision-making with those of the digital computer for rapid computation."[71] For Lockwood, the ideal worker in the task of data reduction was not the able field operator or the diligent scaling "girl"; it was the cyborg.

Notions of feedback and auto-correction had been central to the earliest work in cybernetics.[72] As Peter Galison has noted in his examination of Norbert Wiener's research, "self-correction is *exactly* what Wiener's machines did. Indeed, in every piece of his writing on cybernetics, from the first technical exposition of the AA predictor before Pearl Harbor up through his essays of the 1960s on science and society, Weiner put feedback in the foreground, returning again and again to the torpedo, the guided missile, and his antiaircraft director."[73]

Lockwood's innovation built indirectly on the servo-mechanical world inhabited by anti-aircraft detectors and self-guided torpedoes. It was built

Figure 5.5
The CDC 3200 System. The computer system was the basis of Lockwood's data-reduction scheme. Using a cathode-ray tube and a light pen, (female) human data processors could scale the ionograms directly on the computer display. A larger central computer then generated the resulting $N(h)$ profile as well as its associated ionogram for comparison with the original record. Source: Defence Research Telecommunications Establishment, "Annual Report of the Defence Research Telecommunications Establishment: 1966," DRTE Report No. 1192-U (Ottawa: Defence Research Board, 1968). Reproduced with the permission of the Ministry of Industry, 2016.

around Control Data Corporation's 3200 System (CDC 3200)—a FORTRAN-based computer system introduced in May 1964 and used mainly for scientific data processing but also for satellite tracking and telemetry. (See figure 5.5.) The set-up was simple: two computers (one small, one large), a cathode-ray tube, and a human operator. Its basic operation was also straightforward. The ionographic data passed from magnetic tape through a special-purpose analog-to-digital converter and then to the small computer, which used the cathode-ray tube to maintain a continuous display of the ionogram and some additional information. Using a light pen, the operator then scaled the ionogram directly on the display. The small computer, designed to select a series of data points within a margin centered on the scaled point, used the measured amplitude of the various points and traces to select the data it would use in the computation. The time code and the selected data points were then transmitted to the larger computer, which derived the $N(h)$ profile along with the ionogram that would result from such a profile. As Lockwood explained, the operator then "compares the remaining traces of the original and computed ionogram via the display, and either verifies the scaling or rescales the ionogram until the desired agreement has been achieved." "The data," Lockwood reasoned, "can then be interpreted in one of the possible ways, fed into the computer and the computer used to compute the consequences of such an interpretation. The calculated information can then be displayed so that the operator can look for consistency between the calculated and observed data, and can reinterpret, if necessary, until a satisfactory solution is obtained."[74]

The system had obvious benefits. For one, it worked directly with tape data and therefore preserved information, such as the electrical amplitude of radio echoes, which was lost in conversions to photographic film.[75] Its increased productivity promised to carry forward polar investigations by generating $N(h)$ profiles from "complex ionograms which otherwise could not be used for this purpose without great labour"—seemingly a boon for DRTE. For Lockwood, though, its primary virtue was "its capability of reducing errors in interpretation and scaling by providing facilities for rapidly checking the consistency of the results and for immediate rescaling if required."[76] Through feedback and automation, Lockwood's scheme proposed to collapse the entire reduction process—interpretation, scaling, computation, and verification—into a single iterative and self-consistent operation.

Just below the rhetoric of efficiency and accuracy, the system had ironic effects. Scaling the virtual ionogram was seemingly identical to the desktop procedure carried out at the remote field stations. Whether on a CRT or on a projection table, interpretation and measurement conformed to the same rules laid out in manuals and training seminars. However, from the world of circuits and software—the other side of the computer display—a quite different vision emerged. Here the analyst's careful measurements stood merely as markers. By design, Lockwood's system was not compelled to accept the operator's choice of scaled values. Rather, the computer selected a series of data points within a margin centered on the scaled point. Using the corresponding information about radio echo amplitude for these coordinates, the computer chose a point (or multiple points) corresponding to the strongest signals, which it then passed on for computation.

The underlying supposition was that the strongest echoes corresponded to the most direct (and therefore the most vertical) propagation paths. The effect was to exclude from consideration the oblique echoes characteristic of the turbulent high-latitude ionosphere. Furthermore, the digitization of the ionogram (a prerequisite for the heavy computational needs of the system), combined with the limited capacity of small computers, meant that the gray-scale quality of photographic records was destroyed. The virtual records instead appeared in two-tone, green and black, with the result that "the details of the weaker traces are lost on the digitized ionogram."[77] These weaker traces and the phenomena they represented had fuelled many of the early investigations of the Radio Physics Laboratory. In ostensibly eliminating the two main difficulties in the analysis of high-latitude ionograms—erroneous interpretation and labor intensity—the system also eliminated many of the very traces of the polar ionosphere that ionospheric physicists at DRTE had spent so much effort attempting to legitimize and explain. For Lockwood and his cyborgs, it was not merely the electronic and digital processes of the electronic system that receded into the domain of black boxes; it was the more nuanced characteristics of the high-latitude ionosphere itself and the machinic order it helped explain.

Quite apart from its implications for the ionogram, the system had a number of additional effects. One of these was that it shifted the imaginative work of ionographic interpretation from the data analyst to the computer engineer. As Lockwood recognized, "the sophistication of the processing of the ionogram is limited only by the imagination of the *programmer*."[78] In this case, the system placed this imaginative capacity firmly in the hands of NASA, since John Jackson, NASA's point man for the Alouette project, provided the $N(h)$ reduction program for Lockwood's system. In doing so, the scheme shifted important aspects of data reduction out of the hands of the Canadian human interpreters and backward, through the machinery, to the American software and hardware designers. As Lockwood explained, "the use of programmed criteria requires the designer," rather than the data analyst or scientists, "to make a priori decisions on the data." What had once been the idealizing approximations of ionospheric scientists—assumptions about single-valued distribution functions (that made the density problem solvable), about electron collision frequencies equal to zero (an assumption violated in the E layer), and about the validity of geometrical optics (invalid at the top of the transmission path)—were now hard-wired into circuit boards.[79]

Whereas the move of data analysis from field station to laboratory arguably involved a reconfiguration of skills, the new system was overtly *de*-skilling in nature. The combination of computers and humans had been used in other fields in which pattern recognition and huge data sets were involved—for example, in Luis Alvarez's system for analyzing bubble-chamber images

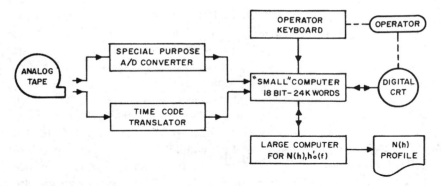

Figure 5.6
A schematic diagram of Lockwood's automated data-reduction system. The symbols used here are taken from the iconography of computer processing. The rectangular boxes represent "computer operations." The human operator is reduced to a "keying operation." © 2016 IEEE. Reprinted, with permission, from G. E. K. Lockwood, "A Computer-Aided System for Scaling Topside Ionograms," *Proceedings of the IEEE* 57 (1969).

at Berkeley. Like Lockwood, Alvarez saw computers as a valuable aid to, though not a replacement for, the unique human abilities of pattern recognition.[80] For Alvarez, the process was potentially iterative, with records undergoing an initial scan, measurement, and computer analysis in which some interpretations were rejected as faulty and sent back for further analysis. But Lockwood's integration was more complete than Alvarez's, blurring more completely the line between human operator and machine. Alvarez saw a system in which human intervention was summoned forth when machines could go no further. Lockwood's was a continuously interactive relationship, a quality captured in his own representation of the system. Using the emerging visual lexicon of computer operations, Lockwood depicted (perhaps for the first time in the history of DRTE) human operators alongside machines in the production and analysis of ionograms. (See figure 5.6.) It was an operationalization in which the human operator was reduced, through her oval signifier, to a "keying operation." As such, the diagram stood for the twofold integration of humans with machines: their physical integration within the iterative loop of analysis and error correction; and their secondary representational assimilation, a recasting of humans in the language and iconography of computer engineering.

In his 1969 brief detailing the system, Lockwood explained the *de facto* training of operators in the "proper" scaling of records through the use of his hardware: "One might expect that, for complex ionograms, several scalings would be required to obtain self-consistency. However, because of his [sic] learning ability, after a short time the operator can usually scale an ionogram correctly on the first or second attempt. With practice, about two minutes are required to scale an ionogram. This includes the thirty seconds needed to digitize the Alouette II ionogram, and the fifteen second wait for the calculated results from the larger computer."[81] The correctness of this scaling was now contained in circuits and in strings of computer code. Lockwood's system transformed the role of data processor, eliminating the long-established practice of using sequences of ionograms to interpret specific phenomena (a practice captured in Meek's f-plot) in favor of the mechanical comparison of original "true ionograms" and the progressively less erroneous products of human action. Here, we are a long way from Leroy Nelms and his imaginative analysts. We are also far along the path by which the long-standing, archetypal "Canadian" traces of the turbulent arctic ionogram were progressively effaced from ionospheric research. The automation that drove this erasure would also drive attempts to circumvent the ionogram and the problematic natural order it helped catalog by creating systems that adapted instantaneously to the turbulence of the upper atmosphere or that sought out alternative natural orders that supported survivable communication technologies rather than thwarting them.

Conclusion

Taken together, the systems for mass production and mass analysis of ionograms re-centered the synoptic project in ways that not even DRTE researchers had imagined possible. The initial concerns about the satellite's reliability, and the continued anxieties about the growing flood of ionograms, ultimately integrated the spacecraft into an expanding infrastructure of tracking and telemetry, driving Franklin's mass-production project, Nelms' imaginative analysis for ionospheric profiles, and Lockwood's scheme for cybernetic analysis. The success of the satellite and of the data-reduction and data-analysis techniques made it possible for DRTE's subsequent satellites (the Alouette-ISIS program) to essentially replace space missions that NASA had envisioned undertaking and encouraged many of that organization's ionospheric scientists to build their research programs around DRTE's efforts. By the time Lockwood was contemplating his cybernetic operators, the Alouette-ISIS program had developed its own telemetry networks independent of NASA's.[82]

That reorientation signals how Alouette and its successors offered an alternative to the major axes of American space diplomacy in the 1960s. Walter McDougall has identified how the US pursued four main objectives through its space programs: protection of American military space programs, cooperation with the Soviet Union in space science and arms control in space, cooperation and competition with European allies, and organization for communications satellites.[83] Cooperation with Canada formed a fifth axis. The Alouette project was not fraught with the same tensions as European cooperation, where US policy resembled the attitude taken toward the Soviet Union: cooperation on scientific matters but circumspection when it came to engineering.[84] Instead, DRTE enjoyed certain exceptional freedoms and used that special status to its advantage. The Alouette project pointed to the role satellite records could play in reorienting contemporary and historical relations in the practice of global science. It also pointed to the role satellites would play in circumventing the problems of natural and machinic orders that had launched DRTE's research in the first place.

6 A Natural History of Survivable Communications

Sometime in 1957, two representatives of the US National Security Agency landed in Churchill, Manitoba, a once-fortified fur-trading post on the western shores of Hudson's Bay. Their ultimate destination was a corrugated steel structure about 3 kilometers west of the airport—a building raised a meter above the ground on stilts (to avoid melting the permafrost) and encircled by three huge arrays of rhombic antennas pointing in all directions.[1] The structure, the largest of its kind in Canada, fed raw signals intelligence directly back to the NSA's Canadian equivalent—the Communications Branch of the National Research Council. The Communications Branch had drawn together various wartime signals intelligence units, including the Operational Intelligence Center, into an organization so secret that its existence would not be officially recognized for another two decades.[2] Exploiting Canada's geographic proximity to the Soviet Union, it intercepted Soviet northern and far eastern communications and fed them directly to the NSA and the British Government Communications Headquarters (the organization that had been responsible for code-breaking operations at Bletchley Park during World War II).[3] Theoretically, the location of the Churchill station and its distance from the electrical noise of temperate and tropical thunderstorms made it an ideal spy nest and an indispensable node in the Signals Intelligence (SIGINT) operations of the West. But its location inside the zone of maximum auroral activity, which had made it a center for research on the disturbed ionosphere, often wreaked havoc on its eavesdropping operations. The NSA representatives, sent specifically to judge Churchill's potential for capturing telemetry transmissions from Soviet ICBMs, arrived in the middle of one of the large-scale short-wave absorption events that originated in the auroral zone, circling Earth with the sun for several days. In the approximately 48 hours they spent there, occasionally playing liar's dice with the station officers, they heard nothing but static.

One of those men was Nathaniel Gerson, successfully and clandestinely courted by the NSA at a conference on ionospheric winds the year before. Gerson, trained in ionospheric physics, would spend the better part of the next four decades as one of the agency's leading intelligence operatives. He had already experienced auroral absorption events firsthand in the late 1940s at Baker Lake, Normal Wells, and Churchill as he criss-crossed the Canadian North for a US Air Force project on long-range navigation. Since then, he had risen to the position of secretary of the American IGY committee and had expanded his interest to signals intelligence of all kinds, especially the errant signals used for ballistic missile detection. The disruptions at Churchill inspired Gerson to move the station's operations as far north as possible, to Alert on the northern tip of Canada's Ellesmere Island—the northernmost permanently inhabited point in the world, closer to Moscow than to Ottawa.[4] The move would avoid auroral absorption while giving the West knowledge that the Soviets had obtained from temporary experimental sites on the Arctic Ocean ice pack. In the years to come, Gerson would experiment with sporadic ionization and chemical bombs detonated in the atmosphere to reflect Soviet telemetry signals to Western listening posts. When these fell short, he shifted his interest to reflections from the Moon. To combat parallel Soviet efforts, he envisioned laying a series of trans-Arctic cables snaking from the arctic shores of North America to Northern Europe and the Pacific. He tied those projects to the emblematic technologies of the Cold War, picturing nuclear reactors tethered to ice floes and powering an enormous mesh of fast, adaptable short-wave stations that would reroute radio-communications around ionospheric disturbances. The NSA considered his imaginative reports so sensitive that their unauthorized disclosure would cause "exceptionally grave damage" to national security; material so secret that the intelligence community classified it according to their most clandestine category of raw intelligence, denoted by a single code word: DAUNT in the late 1950s and, eventually, UMBRA—or "shadow."[5]

Gerson occupies the deepest shadows of DRTE's work, just as the clandestine details of that work have, until now, occupied the shadows of these pages. By the time he visited Churchill for the NSA in 1957, Gerson had known Frank Davies for a decade and considered him a mentor. He was on a first-name basis with Jack Meek, who would welcome him to "the club" during a 1958 visit to the headquarters of the Defence Research Board.[6] In his time with the Air Force, he had sponsored auroral investigations at Saskatoon, where a number of DRTE researchers were seconded; the research projects and reorientations of his own laboratory on the outskirts of the MIT campus often shadowed those of DRTE; he shaped space physics curricula at universities across Canada (including the University of Western Ontario, where John Chapman had been trained); he set out the topics and training that would guide generations of Canadian atmospheric physicists whose projects ultimately fed back into his own ultra-secret work. His connections and collaborations with DRTE formed part of a vast network of secret relationships, forged in the early years of the Cold War, between civilian scientists, federal agencies, and the intelligence communities of the United States and Canada—relationships designed for global surveillance and the prospect of nuclear war.[7]

Gerson's vision of electronic circumpolar surveillance and communications forms an organizing principle for this chapter. DRTE's research focused publicly, almost obsessively, on the ionogram and its role in mapping and predicting the high-latitude ionosphere and charting the machinic order of high-frequency radio. But that research included smaller, more opaque, less visible projects that evolved in parallel, and sometimes in tension with that very public objective. As Davies himself recognized openly in the early 1950s, the picture of the upper atmosphere he had sketched in 1943 had been far too simple. Ten years later, the ionospheric winds were still poorly understood; so were the mechanisms of solar-terrestrial physics that seemed to govern the creation and dynamics of the ionospheric layers. Frustrated and facing a developing Soviet nuclear threat, his laboratories turned to projects running quietly in parallel with the sounding and prediction programs—projects that sought to turn the atmospheric anomalies that had plagued shortwave radio into resources for reliable communications and simulated war, or that sought to build machines capable of adapting dynamically and immediately to the changing ionosphere, avoiding the ionogram altogether.

Seen through a system of documentation and publication within Canada that was designed precisely to fragment their coherence, to obscure an overall picture of their activities, those projects fit awkwardly into the story we have followed thus far. But seen through the comprehensive vision of one person who helped initiate them, who sought to turn the high-latitude natural order into an instrument of reliable communications and Cold War spycraft, they gain a startling coherence. For Gerson, the Canadian North was only part of a much broader trans-Arctic and ultimately global system of communications, an emerging and evolving machinic order of the polar regions that needed to be secured against failure, jamming, and Soviet espionage, and exploited for Western communications and intelligence. His interests positioned specific, characteristic elements of DRTE's archetypal natural order as elements in that more secretive Arctic and polar order of the Cold War.

This chapter examines two projects that functioned under that rubric, and that formed part of a wider history of survivable communications in the early Cold War. The first sought to use meteor trails as reflection surfaces for short-wave radio transmissions, bypassing the traditional ionospheric layers to create a communications system that could withstand not only auroral storms but also their human-made analogues—nuclear detonations. The second attempted to build machines that would automatically probe the ionosphere, choosing the best communications frequency for a given circuit with no need for ionograms. Each system embodied concepts that would be central to the larger project of engineering communications capable of surviving nuclear attacks: in the first case, store-and-forward communications, which logged information at network nodes for onward transmission when reliable connections were made; in the second, real-time channel switching that immediately and automatically determined optimal transmission circuits between two points. Each would form part of an evolving vision of distributed communications that included Paul Baran's concept of packet switching, often seen as the origin of the Internet.[8]

This chapter, however, shows how the history of those concepts is, in part, a *natural* history: a history of the natural orders that helped generate these systems and their central concepts. The history of survivable communications—communications that could not be allowed to fail—is partly a history of the relationship of specific machines to the natural orders delineated during the Cold War. It is also partly a history of disappearance. Both projects described here were conceived out of the failure of ionospheric prediction, growing out of an attempt to develop systems that could adapt to the turbulent and changing upper atmosphere in real-time, either circumventing the complex and unpredictable ionosphere altogether or else making its turbulence essentially invisible through automation. Together, the two systems and their fate signaled the decline of the ionogram as a tool linking natural and machinic orders, creating the conditions for the final abandonment of the Northern radio project, which we explore in the final chapter.

Sporadic Phenomena

The theoretical understanding of short-wave radio-communications, the one that guided DRTE's work for much of our story, had been founded on the idea of a smooth and uniform ionosphere. Throughout the late 1930s and the 1940s, though, radio research had cataloged anomalous and inhomogeneous sporadic phenomena that interfered with that neat picture of reflected transmissions. As defined in the late 1940s, "sporadic" ionization encompassed two types of phenomena: ionization that was irregular—scattered or isolated spatially, or occurring at varying intervals; and phenomena that changed rapidly or erratically. Phenomena of both types had been at the center of DRTE's claim about the unique natural and machinic orders of the Canadian North. Many of the phenomena at the center of DRTE's research—spread echoes, polar spurs, sporadic E ionization—fell into one of the two camps; others, like the aurora, occupied both.

In the early years of the Cold War, it was precisely their status as unpredictable exceptions to the natural order that made sporadic phenomena interesting in an age of imminent attack. The phenomena broke with the regularity of ionospheric physics in ways that gave potential insights into warfare in the nuclear age. After the Soviets tested their first atomic bomb in 1949, "sporadic" atmospheric phenomena were assimilated into even more general "sporadic" events that interfered with communications and detection, including bird migrations that fooled early-warning radars.[9] Once the ionospheric prediction systems began to break down in the early 1950s, though, particularly for high latitudes, the irregular, erratic nature of sporadic phenomena came to stand in as analogues of archetypal but unexpected "artificial" events that might precede or attend a nuclear war. Through them, the machinic orders of the Cold War and the natural order of the upper atmosphere were constantly read back and forth against one another. Meteor trails in the upper atmosphere simulated the wake of ionized gas that would trail ballistic warheads, driving research into radar and optical sensors to detect and track them. Aurora were believed to mirror the visual and electromagnetic effects of a nuclear detonation in the upper atmosphere, disrupting communications but also masking airborne objects (initially airplanes and, later, missiles), either optically or through disruptions to the early-warning radars being woven into plans for continental defense. The attempts throughout the 1950s, particularly by Jack Meek, to incorporate sporadic phenomena into the body of ionospheric research was one sign of that more complex status and valuation of the phenomena. Meek's efforts signaled the more general drive to study and classify sporadic phenomena as a way of predicting and compensating for them, but also of possibly turning them to advantage. As such, the phenomena were both highly public and intensely classified. They formed part of the communications disruptions with which DRTE openly identified itself, and of the highly secretive orders of surveillance and simulated war in which they were implicated.

The dual status of aurora in particular had drawn the attention of Frank Davies' friend Balfour Currie, whose prewar and wartime auroral investigations at the University of Saskatchewan had turned it into a global center for research on the subject after the war. Its success attracted the attention of the United States Air Force in the person of Nathaniel Gerson, the newly appointed head of the Cambridge Research Laboratory's Electronic Propagation Section. Gerson had just completed a master's thesis on ionospheric ionization at New York University while working for the Air Force, and had become interested in what he considered possible puzzles in radio propagation and fading in the polar ionosphere. He had met Frank Davies in Portage La Prairie in 1946 while inspecting long-range radio navigation equipment at the Canadian air force base there. Davies had arrived as part of his work for the Canadian Radio Wave Propagation Committee to install an ionosonde in the same room where Gerson was working. Gerson would later operate the device almost continuously during the Perseid meteor shower, looking in vain for echoes from ionized meteor trails—a topic that would later form part of his sweeping vision for polar communications. Both men shared a fascination with geophysical anomalies and "sporadic" phenomena of all kinds, and their interests extended to questions of motivating researchers and administering laboratories. Gerson would later intimate that Davies was one of two people from whom he had learned everything he ever knew about research administration.[10]

Gerson had deep affinities with DRTE's approach to ionospheric communications. Like Davies' group, he had quickly become disillusioned with the heavy emphasis on statistical predictions for short-wave radio, believing they placed too much emphasis on diurnal and seasonal averages. For Gerson, this was like asking pilots to use climate data rather than current weather patterns when planning a flight path.[11] From 1948 to 1955, as chief of the Ionospheric Physics Laboratory at the Air Force Cambridge Research Laboratories (AFCRL) in Massachusetts, he had shifted the focus of radio propagation research "away from dust-gathering statistical studies toward original research on emerging new theories about the physics and dynamics of the ionosphere." Like Davies and his colleagues at RPL, he began exploring the mechanisms behind ionospheric formation and dynamics, changing the name of the laboratory from Electromagnetic Propagation to Ionospheric Physics.[12] Like Davies, he focused heavily on identifying "precursors"—solar flares, magnetic disturbances, and other phenomena that formed the first links in a chain of events that culminated in radio disruptions or complete blackouts.[13]

Almost immediately after taking lead of his laboratory, Gerson enlisted the existing program in auroral physics at Saskatchewan, openly funding spectroscopic analyses and observations to help in target identification, sending an early-warning radar of the same type that had detected the approaching Japanese attack at Pearl Harbor, and encouraging the Defence Research Board to fund radar work there.[14] At the same time, he arranged additional university contracts: radio meteorology at McGill, geomagnetics at the University of Toronto (under Tuzo Wilson), and auroral physics at the University of Western Ontario, where John Chapman would earn his doctorate in atmospheric physics and where, in the summer of 1951, Gerson organized an influential international conference on auroral research. The institutional arrangements briefly upset Frank Davies, who felt that Gerson was pouring too much money into too many Canadian universities, creating unrealistic expectations.[15] The program at Saskatchewan, though, would an important training ground for seconded DRTE scientists pursuing research projects and advanced degrees. Jack Meek would carry out his careful correlations of auroral, magnetic, and ionographic observations on Gerson's equipment, searching for the same "precursors" that signaled oncoming disruptions. And throughout the 1950s, as their faith in short-wave predictions waned, both institutions developed projects that sought to exploit a range of sporadic phenomena—from aurora to meteors—shifting them from communications obstacles to surveillance and communications instruments.

By 1960, Gerson had enrolled sporadic phenomena in questions so secret, their very existence was denied. In 1956 he had been invited to present a paper on ionospheric winds at Arlington Hall Station, a former private girls' school that had served as the wartime headquarters for Army cryptographers and, seemingly unbeknownst to Gerson, for NSA signals intelligence since 1952. After his talk, the NSA moved to recruit Gerson. From then on, he would identify himself as a consultant for various government agencies, including ARPA, to cover both his intelligence activities and the existence of the NSA.[16] The timing of his recruitment was no accident. While serving as secretary of the American national IGY committee, he was approached by the NSA just as rocket and satellite experiments were getting underway. He worked in the NSA's Research, Engineering, Mathematics, and Physics (REMP) division, initially planning the curriculum for its College Park Facility and working on laboratory studies that simulated ionospheric conditions. His work increasingly turned toward applying his geophysical research for the Air Force toward intercepts, the "front end" of signals intelligence systems. Eventually he would develop techniques for direction finding, radio fingerprinting, and spread-frequency communications, advising the NSA on how to pick up even the weakest signals.[17] It was in that capacity that he advised CBNRC in 1957 at Churchill and then at Alert.

In the next five years, Gerson would develop a series of remarkable projects that sought to exploit geophysical anomalies for the purpose of signals intelligence, supporting propagation over extended distances and allowing the detection of missiles and satellites beyond horizon.[18] He explored the possibility of antipodal propagation based on "whispering galleries" like those in Washington's Statuary Hall and in the Louvre—elliptical rooms that focused sound waves in such a way that hushed sounds produced at one focal point could be clearly heard across the room at the other. Gerson identified "natural" whispering galleries, including the space between the ionosphere and the earth's surface, where radio reception was strongest at two points—in the vicinity of the transmitter; and at its antipode, precisely halfway around the world, suggesting a mechanism for intercepting Soviet transmissions in the opposite hemisphere.[19] Later in the decade, he would pursue artificial means for capturing the transmissions—artificial electron clouds, orbiting dipoles (essentially a cloud of metallic needles), satellites, and even nuclear detonations.[20] But throughout most of the 1960s he focused on naturally occurring atmospheric phenomena: sporadic E clouds that reflected Soviet signals to the listening post at Alert, high solar activity that raised the frequency limits of ionospheric propagation to 40–50 MHz for distances of 4,000 kilometers, auroral ionization that directed ICBM telemetry to North America, and magnetic channeling and scattering from meteor trails.[21]

The problem of intercepting signals using sporadic phenomena formed precisely one half of a communications problem. The idea of scattering itself stretched back to the early years of the twentieth century, when Arthur Kennelly suggested the possibility that radio waves might be scattered as well as reflected by the ionosphere. In 1950, a conference at MIT attended by Lloyd Berkner and Henry Booker suggested that the phenomena might be used as the basis of a communication mechanism. (See figure 6.1.) In the same way that water droplets in fog might scatter a light beam, inhomogeneities in the ionosphere were believed to scatter radio waves in ways that might be exploited not only for signals intelligence but also for communication circuits using a powerful transmitter and sensitive receivers. The signals would be weak, limiting the transmission distance to where the scattering point met the horizon (as seen from the receiver). But this kind of scattering had the advantage of being reliable under otherwise disturbed conditions. Auroral storms, sudden ionospheric disturbances from solar flares, and polar blackouts—precisely the phenomena that disrupted Northern communications—generally strengthened scattered signals.[22] Experiments would focus on scattering from the ionosphere, but unexpected

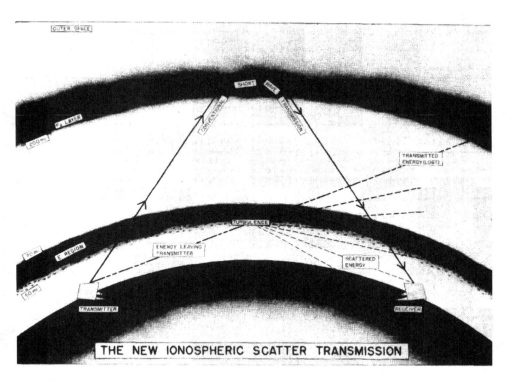

Figure 6.1
"The new ionospheric scatter transmission." Normal short-wave communications use radio waves reflected from the outer layers of the ionosphere. Ionospheric scatter instead used high-power transmissions reflected from turbulence in the lower portions of the ionosphere. Source: National Bureau of Standards, "Forward Scatter of Radio Waves," *Navigation* 5, no. 2 (1956): 107–113. Reprinted with permission of John Wiley & Sons, Inc.

results from other inhomogeneities in the region, particularly meteor trails, ultimately launched research projects designed to bypass the ionosphere altogether.

Meteors

In the middle decades of the twentieth century, meteors had undergone a dramatic transformation from radar nuisance to communications medium. Ionospheric researchers working as early as the 1930s had hypothesized that ionized trails left by meteors would cause density fluctuations in the ionosphere. During the V2 barrage of London, false alarms from the British Army's gun-laying radars were attributed to these kinds of fluctuations.[23] After the war, a succession of British researchers, including Edward Appleton, turned their attention to using reflected radio waves to observe meteors.[24] The largest group was at Jodrell Bank Experimental Station, set up in the botanical gardens of the University of Manchester. While searching for cosmic rays there in late 1945, Patrick Blackett, the left-wing scientist who had served as director of the British Admiralty's Operational Research during the war, along with his protégé Bernard Lovell, discovered the echoes of meteor trails using former army radar equipment.[25] During the Perseid shower in July and August 1946, as Gerson watched Davies' ionosonde for meteor echoes, Lovell investigated the possibility of detecting them on radar. In October he succeeded in registering thousands per hour, giving the impression of "a great array of rockets coming towards one." The next year, immersed in meteor studies, he and his colleagues observed echoes from aurora.[26]

At the core of Lovell's meteor research had been a question about origins. Meteors that entered the atmosphere as part of a shower often seemed to originate from a single point, called a "radiant." But the vast majority, known as "sporadic" meteors, seemed to arrive irregularly from all directions raising questions about whether they originated within the solar system or came from interstellar space. The shape of the meteors' paths could be used to settle the issue—parabolic or elliptical paths indicated they orbited the Sun; hyperbolic paths indicated their origin was interstellar. Between 1948 and 1950, Lovell and two colleagues undertook a careful study of radar echoes from meteors and identified their paths as parabolic. At the Canadian National Research Council, Donald McKinley used their techniques and a twin of the auroral radar at Saskatchewan, also supplied by Gerson, to launch his own study of sporadic meteors. By October 1950, he had combined radar, photographic, and visual observations of 10,933 sporadic meteors, and reached the same conclusion about their origins.[27] In that same year, McKinley became the first to suggest that these sporadic meteors might be used to communicate over considerable distances.

The RPL project on meteor communications came out of an experiment that unintentionally confirmed that speculation. In early 1951, researchers from MIT's Lincoln Laboratory had set out to investigate the possibility of long-distance communications by scattering powerful VHF signals from sporadic phenomena in the lower ionosphere. They set up a continuous transmission between Cedar Rapids, Iowa and a receiver at the 450-acre Central Radio Propagation Laboratory in Sterling, Virginia, roughly 1,200 kilometers away. RPL researchers in Ottawa picked up the distant signal and began monitoring the transmission. Because of the distance, they concluded that the signal was being reflected from meteor trails. Peter Forsyth, a McGill PhD and RPL physicist fresh from a year-long secondment to Saskatchewan, inherited their experiment the following year. Forsyth had observed the first radar aurora at Saskatoon in 1949, using the early-warning radar Gerson had furnished.[28] Gerson would continue to support Forsyth's research throughout his time at RPL and after Forsyth had left for the University of Western Ontario and Gerson himself had turned to signals intelligence for the NSA. Forsyth soon realized that the seemingly continuous signal was actually intermittent, composed of a huge number of discrete signal reflections from individual meteor trails. Researchers working for the National Bureau of Standards had received the same intermittent signal at Sterling alongside the direct, continuous signal and considered it a nuisance to be eliminated.[29] With nothing *but* the meteor interference to work with, Forsyth and his colleague Eric Vogan decided to use the interference as the basis of a communication system.

Growing out of the radar aurora studies at Saskatoon, they seized on a suggestion, made twenty-five years earlier, that radio waves scattering from inhomogeneities in the ionosphere might provide a way of communicating with points that were otherwise inaccessible.[30] The aurora, they noted, provided one source for these inhomogeneities. Their own research at Saskatoon explained that auroral radar echoes came from small, intensely ionized regions in the aurora.[31] Another source, they asserted, were the roughly ten billion meteoric particles that bombarded Earth's atmosphere each day. As these objects entered the thermosphere (the denser portions of the atmosphere between 80 and 120 kilometers), collisions with surrounding air molecules heated them, producing visible light along with a trail of free electrons, sometimes several kilometers long, in their wake, like the high-altitude contrail of a jet plane. Because of the small size of most of the particles, only two or three meteors per hour might be visible from Earth's surface as "shooting stars." But hundreds of these traces were "visible" to VHF radio equipment.[32] If used as reflection surfaces, these sporadic phenomena might furnish the polar regions with a communication system that would remain reliable, even in the face of atmospheric nuclear detonations, by avoiding the ionosphere altogether.

Vogan and Forsyth began piecing together their understanding of a natural order in which meteors now figured as protagonists. Earth's atmosphere, they explained in concise, often poetic prose, was continuously bombarded by small, dust-like particles. The billions of these particles that showered the planet every day weighed a total of only about a ton. Larger objects, which formed meteorites that produced fireballs and struck Earth, were rare, making them uninteresting for communication purposes. Drawing on early British and Canadian work on sporadic meteors, they explained that the meteors themselves seemed to have their origins in interplanetary (rather than interstellar) space. Their most important properties were essentially random, impossible to predict except statistically. They varied greatly in size, number, velocity, and orientation, entering the atmosphere from all directions; they varied with the seasons and with the time of day. At local dusk, particles would shower overhead as "an observer on the Earth's surface is being carried headlong into the meteoric particles"; at dawn, the barrage would wane as the particles were forced to overcome Earth's rotational velocity.[33] As the particles entered the denser portions of the ionosphere, atmospheric molecules heated them, evaporating atoms on their surface; those atoms then struck more air molecules at thermal velocities, ionizing a number of the atoms and creating an ionization trail that was opaque to radio waves. The combined action of billions of these meteoric particles created a complex and dynamic three-dimensional lattice of ionized trails criss-crossing the upper atmosphere, individually coming into existence and fading away; a constantly changing skein of possible reflection surfaces.

The main difficulty in using this lattice for communications was that, for any two points on the planet's surface, not all trails would support communications. A trail's ability to reflect transmissions from one specific point to another was a complex function of three-dimensional position, orientation, and ionization. A research group at Stanford, funded by the AFCRL, had been investigating these questions in parallel.[34] Drawing on their research, Forsyth and Vogan quickly realized that their system would have to operate in bursts. By spreading communications over successive meteor trails, they set out to develop an intermittent, "meteor-burst" system that sent portions of prerecorded messages at high speed when a suitable trail existed and closed down transmission when it disappeared. Given the impossibility of predicting the occurrence of suitable trails, the difficulty lay in devising a machine that would evaluate the multitude of possible trails and select the appropriate ones.

Forsyth and Vogan's machine operationalized that problem. They created a kind of call-and-response system that only started transmissions when a closed circuit was created. For a given communications circuit, each terminal would be outfitted with a transceiver continuously transmitting a carrier wave. The constantly forming and disappearing meteor trails would reflect the transmission, scattering it in various directions, the bulk of them missing the receiving station altogether. Eventually a meteor trail with the right orientation, position, and density would appear and reflect the carrier wave to the receiver. The receiving station would then detect the carrier signal from the transmitter and send back an acknowledgment (in the form of a modulated tone). If the transmitter received the modulated tone with sufficient strength, it would begin high-speed transmission of the prerecorded message using magnetic tape. The transmission would continue as long as the acknowledgment persisted. When the tone stopped or was lost, the transmitter would end the transmission and wait to be triggered again by another acknowledgment. Since the burst communication only used dense ionization trails, it could operate at lower power than a system designed to use weaker trails more frequently.[35] The system would only transmit a small fraction of the time, while it waited for meteor trails with the proper characteristics, but its high volume would make up for the low duty cycle.

Although the system operationalized the problem of discriminating between meteor trails, reducing it to a problem of signal detection, Forsyth and Vogan understood their machine in less operational terms. Drawing on the language of radar that had surrounded him at Saskatchewan, Forsyth repeatedly presented the machine as "detecting" the myriad trails continually forming and disappearing and then "selecting" a suitable trail through gating circuitry—the way an automated radar might detect and discriminate between incoming targets. In reality, of course, the machines were blind to the mechanism by which the radio wave traveled between them; they would establish a circuit as long as they sensed radio signals above a certain threshold intensity determined by gating circuits whose parameters had been set by the researchers. But the impression Forsyth and Vogan gave was of a machinic order of VHF transceivers specifically attuned to a dynamic natural order centered on meteors, responding almost instantaneously to the transient, sporadic phenomena. (See figure 6.2.) Their description of the system mixed the natural and machinic effects, reading the natural order of sporadic meteors into the behavior of the machine. A specific group of machinic effects—in this case the burst or silence of the teletype—was an indication of otherwise unobserved meteor trails in the upper atmosphere, just as the static and malfunctioning of radios or the smeared traces on ionograms had indicated ionospheric disturbances in the decade before.

They called the system JANET, after Janus, the two-faced Roman god of transitions who presided over gates, doorways, and passages, but also

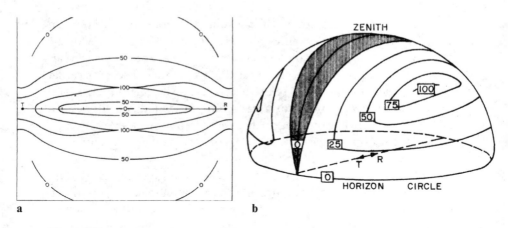

Figure 6.2
(a) The spatial distribution of useful meteor trails. (b) The distribution of useful meteor radiants, shown using contours on the celestial hemisphere. © 2016 IEEE. Reprinted, with permission, from P. A. Forsyth et al., "The Principles of JANET: A Meteor-Burst Communication System." *Proceedings of the IRE* 45, no. 12 (1957): 1642–1657.

over the beginning and ending of conflict, and therefore over war and peace. To test the principles of the system in the summer of 1953, roughly two months before the Soviet Union tested its first hydrogen bomb, they built experimental equipment with the help of a number of other researchers, including Del Hansen, who had worked the RPL Mobile Observatory and now headed the technical services wing. Using a railway line running along a pier to the grain elevators in Port Arthur (now Thunder Bay), they mounted the equipment on a railway pushcart and guyed the antennas to the railway tracks. Before each of their trials, they pushed the cart (stored in one of the grain elevators) out to the antennas and connected them, then listened for the modulation tones that signaled a closed circuit with Ottawa, 1,050 kilometers away.[36] Electrical noise from motorboats carrying curious onlookers interfered initially with the signals and they had to search for the precise frequency being transmitted. After several days, the equipment detected the carrier signal and automatically sent out the modulated tone that was then received in Ottawa.

The success of the test led them to build a prototype of the communication system itself—the first truly operational meteor-scatter communication system.[37] After modifying the machines to try to reduce the error rate and refine the gating circuits, it was tested late in the winter of 1954. "The

MORE THAN A QUARTER OF A CENTURY AGO,
ECKERSLEY POINTED OUT THAT SCATTERING FROM
INHOMOGENEITIES IN THE IONOSPHERIC IONIZATION
COULD BE AN IMPORTANT FACTOR IN THE PROPAGATION
OF RADIO WAVES AND COULD INDEED PROVIDE A MEANS
OF COMMUNICATING WITH STATIONS WHICH OTHERWISE
WOULD BE INACCESSIBLE. SINCE THAT TIME THERE
HAVE BEEN MANY INVESTIGATIONS AIMED AT THE
UNDERSTANDING AND UTILIZATION OF THE VARIOUS
IRREGULARITIES WHICH OCCUR IN THE IONOSPHERE.
THE UNDERLYING HOPE IN MOST OF THESE STUDIES
HAS BEEN THAT SOME MEANS WOULD BE FOUND WHEREBY
AT LEAST THE LOWER PART OF THE VHF BAND COULD
BE USED FOR LONG DISTANCE COMMUNICATION.

(a)

XX   MORE THAN A QUARTER OF A CENTUXXRY XXAGO,
XX
ECKERSLEY POINTED OUT THAT SCATTERING FROM
INHOMOGENEITIES IN THE IONOSPHERIC IONIZATION
XX
COULD BE AN IMPORTANT FXXACTXXOXR IN THXXE PRXXOPAXXGATION
OF RADIO WAVES AND COULD INDEED PROVIDE A MEANS
OF COMMUNICATING WITH STATIONS WHIXXCH OTHERWISE
WOULD BE INACCESSIBLE. SINCE THAT TIME TXXHERE
HAVE BEEN XXMANY INVESTIGATIONSXX AIMED AT THE
UNDERSTANDING AXXXXND UTIXXLXXIXXZATION OF THE VARIOUS
IRREGULARITIES WHICH OCCUR IN XXTHE IONOXXSPHERE.
THE UNDERLYING HOPE IN MOST OF THESE STUDIES
HAS BEEN THAT SOME MEANS WOULD BE FOUND WHEREBY
AT LEAST THE LOWER PART OF THE VHF BAND COULD
BE USEXXD FOR LONG DISTANCEXX CXXOMMUNICATXXION.

(b)

Figure 6.3

A sample message sent through the JANET meteor-burst system. The original message is on the left; the transmitted message is on the right, with XX denoting breaks in the message as the system changed meteor trails. © 2016 IEEE. Reprinted, with permission, from P. A. Forsyth et al. "The Principles of JANET: A Meteor-Burst Communication System." *Proceedings of the IRE* 45, no. 12 (1957): 1642–1657.

system would have delighted Rube Goldberg," Forsyth mused.[38] Two teletype machines were set up six feet apart in a laboratory in Ottawa. Typing on one of them would produce the message a few seconds later on the other, the signal having traveled to Dartmouth, Nova Scotia and back by way of a meteor trail (a relatively inefficient way of communicating across the room). Once transmitted, the message was decoded, printed and compared to the original. (See figure 6.3.)

Along with the research of the Stanford group, JANET would come to form part of an evolving vision of "survivable" communications systems of the late 1950s, designed to continue functioning under focused attacks and heavy damage. The central tenets of those systems were most famously elaborated by Paul Baran of the RAND Corporation in the early 1960s and presented in 1964 as an eleven-part study titled *On Distributed Communications*. The study was written mainly for military systems planners dissatisfied with the reliability and survivability of existing communications networks.[39] Baran's initiative bore a double relationship to the nuclear age. The communications disruptions he envisioned took the form of targeted nuclear attacks. But if the communications that controlled weapons systems could be made more survivable, they would increase the retaliatory capabilities of the West and help deter an assault in the first place, making war less likely. In the early 1960s, the extreme vulnerability of long-distance communications networks made their survival doubtful—a deeply destabilizing situation for Baran.

The key to Baran's system was its distributed nature: the fact that any node in the network was connected to all adjacent stations, meaning the system lacked a hierarchical control center that might be targeted. Distributed systems took many forms in the late 1950s and the early 1960s, as the threat of large-scale nuclear attacks became more pressing. Meteor-burst communications such as those used in JANET were one example of a store-and-forward system that would become popular later in the decade. Although they largely met Baran's interest in "survivability in an unfriendly environment," meteor-communications were "fundamentally unreliable" because useful meteor trails occurred only on a random basis. Baran instead focused on systems with routing flexibility and capacity for growth, always visualizing his network as a grid. If one set of nodes were damaged, automatic, local switching would reroute messages around non-functioning parts of the network. In this way, the system found the "best" surviving paths in a heavily damaged network.[40]

Despite future reservations like Baran's, Canadian officials were initially enthusiastic about JANET. The system had clear benefits for the sparsely populated North, whose need for low-power, reliable point-to-point communications over ranges from 500 to 1,500 kilometers had originally made short-wave communications so attractive. Also, transmissions were highly directional; unlike other forms of radio, the signals could only be received in a relatively small area on the ground, which varied according to the size and orientation of meteors. That property made them inherently secure against electronic eavesdropping of the kind carried out at Churchill. Unlike short-wave, however, the system was not seriously affected by magnetic and auroral storms, or nuclear detonations, but it could be affected by particularly strong atmospheric noise, like local storms, producing errors.[41] In recommending the system, the authors emphasized the universality of meteoric particles and their trails. Because of this universality, their system would be largely immune to the atmospheric disturbances that plagued high-frequency radio as well as to the interference from the crowded high-frequency band. Furthermore, the system was reliable in Canada precisely to the extent that the nation's position in the natural order of interplanetary space was *un*exceptional.

Despite the initial enthusiasm, the system was ultimately abandoned by late 1958—around the time when DRTE's research was redirected toward ICBMs and the proposal for S-27 began to take shape. During later tests between Yellowknife and Edmonton, it was severely affected by a Solar Proton Event and was considered obsolete by the late 1950s.[42] One of its principal weaknesses was that its performance could only be predicted statistically—there was no way of elaborating a physical, dynamical link, a mechanism, that would predict the number, size, orientation, or position of meteors.[43] The same year of its cancellation, Forsyth would leave RPL for a professorship at Saskatchewan. In 1961 he would take up a position as chair of physics at the University of Western Ontario, where Gerson would continue to fund projects in auroral and space physics around these questions.

Adaptable Communications

The simplicity and general reliability of short-wave in most parts of the world had led to its enormous expansion in the middle decades of the twentieth century until it formed the backbone of all long-distance radio communications in the late 1950s, and a backup for all others. Even against competition from other systems like JANET, it maintained its lead. But it was also the most susceptible of all the possible systems to geomagnetic disruptions, solar effects, and particularly interference and jamming.[44]

In 1964, the same year in which the film *Dr. Strangelove* fictionalized a nuclear holocaust arising from a communications failure and Paul Baran published *On Distributed Communications*, Gerson turned to the question of survivable Arctic communications. The difficulties faced by short-wave had focused investigations on two approaches, he felt. The first was on improving the design of communication circuits, irrespective of frequency. Forsyth's meteor-burst communications and the related technique of tropospheric scatter fell into this camp. So did a tentative proposal to lay a submarine cable through the Arctic Ocean linking Europe and North America. (See figure 6.4.) The cable faced the potential of sabotage and damage where it made land-fall, and might be subject to Earth-current storms; laying it was challenging, but not impossible in an age of nuclear-powered icebreakers and submarines.[45]

The second approach focused on developing distributed systems that would reroute short-wave traffic around blackout areas.[46] (See figure 6.5.)

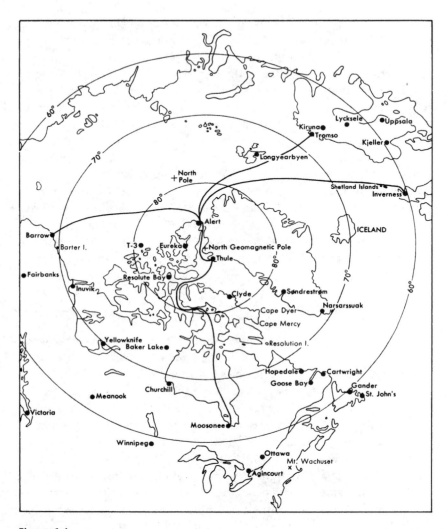

Figure 6.4
Gerson's proposed submarine cable for intra- and inter-arctic communications. The cable would have had terminals at Alert, Thule, Barrow, and Moosonee, connecting it to existing networks in Europe. © 2016 IEEE. Reprinted, with permission, from Nathaniel C. Gerson, "Radio-Wave and Alternative Communications in the Arctic." *Electrical Engineering* 81, no. 5 (1962): 364–366.

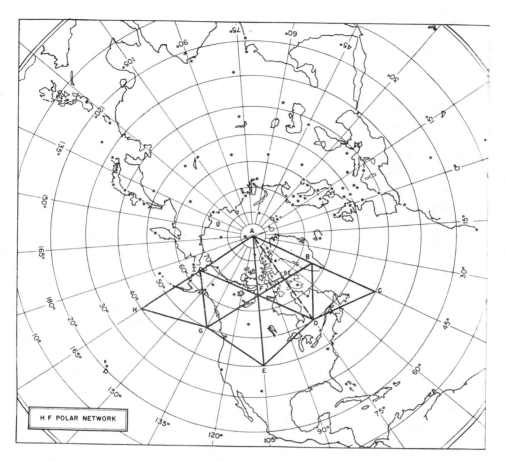

Figure 6.5

"HF Polar Network." Proposed at a NATO conference in Italy in 1963, the network would reroute HF communications around polar disruption events. Source: S. C. Corontini et al., "The Problem of Arctic Communications Following Solar Disturbances," in *The Effect of Disturbances of Solar Origin on Communications; Papers Presented at the Symposium of the Ionospheric Research Committee, AGARD Avionics Panel, Naples, Italy*, ed. George J. Gassman (Pergamon, 1963). Reprinted with permission of Elsevier.

The initial idea came to Gerson's attention at a NATO conference in Italy three years earlier, where researchers working with the support of AFCRL proposed a network of short-wave transmitters and receivers spanning the North American Arctic. In the event of a blackout, radio operators would use statistical data to reroute communications along different frequencies or transmission paths.[47] In his commentary on the proposal, Gerson felt that such a relay system was feasible. But instead of relying on human operators and statistical data, his bête noir since the early days of AFCRL, the system would provide real-time knowledge about open links between nodes in the network. The key to Gerson's plan was a grid of "sensing stations," forming a giant mesh whose edges spanned the Pole and whose individual stations routinely checked communications paths to all other stations in the grid. The stations would be spaced 500 to 1,500 kilometers apart across the Arctic, subarctic, and lower latitudes. They would use short-wave transmitters, receivers, data-storage systems, and switching circuits "collocated" with oblique ionospheric sounders to give instantaneous information about the ionosphere. The originating station would feed its communications into the grid; the grid would use the oblique sounders to automatically switch to open links, forming a communications channel. Gerson was mute on the mechanism that would actually "choose" the best path through the network, but like Baran's system, it privileged two factors: a distributed structure—the fact that each sensing station was connected to all adjacent stations; and automaticity—local switching to automatically reroute messages around non-functioning parts of the network.[48] "Rapid switching techniques at each relay site," Gerson explained, "would permit instant exploitation of any effective openings and permit the flow of information from originator to destinee."[49] To demonstrate the feasibility, Gerson again invoked the nuclear age that made such a network both necessary and possible: "Modern industrialized nations possess the potential and the means of establishing transmitter or receiving sites anywhere in both polar regions, or on ice floes of the Arctic Ocean, with little difficulty and with no great perturbations to their economy. Indeed, the increasing availability of nuclear power plants will permit the actual location of communication stations to be determined solely by technical considerations."[50]

But Gerson's system also differed from the distributed systems of the early 1960s in the way it tied the idea of survivability to a natural order. In this way, it pointed to a broader move in the late 1950s away from the recognized failures of ionospheric predictions and toward a constant real-time surveillance of the ionosphere in an attempt to either circumvent or exploit the peculiarities of high latitudes through automation and adaptation.

Gerson's projection of a transarctic short-wave grid uncannily resembled a much more secret system, developed around the same time, of continuously operating short-wave transmitters and receivers on either side of the Sino-Soviet land mass to detect missile launches from Soviet or Chinese territory.[51] Instead of sensing and adapting to natural disruptions in the ionosphere from polar cap absorption events and magnetic storms, the system monitored the ionosphere for changes with artificial origins. A missile penetrating the region would disrupt the ionization, causing fluctuations in the radio transmissions and alerting Western monitors, first in Aviano, Italy, and ultimately at NORAD. But it also bore an uncanny resemblance to a project at DRTE that would seek to use the same elements—the abandonment of prediction, the use of oblique sounders, the dependence on short-wave, real-time adaptation, and automatic rerouting—to avoid having to deal directly with the vagaries of the high-latitude ionosphere.

That initiative began as a project known as Rampage, which itself grew out of a six-year investigation that started in 1954 and used oblique sounders to investigate the "fundamental limitations" of ionospheric predictions.[52] Oblique sounders yielded their own form of ionograms, less versatile than standard ionograms (since they could not furnish electron density profiles) but more directly applicable to communications since they illuminated actual propagation conditions. Before their introduction in the mid-1950s, researchers had only a vague idea of the actual properties of the short-wave communications channel being used.[53] Other groups had proposed using oblique sounders to study radio propagation between fixed points, but the DRTE investigations were the first to suggest that they be used as a direct aid to communications.[54] The six-year program uncovered apparently insurmountable flaws in even DRTE's complex prediction system, developed specifically for polar regions.[55] Based on these results two groups at DRTE, one under Eldon Warren and the other under Lou Hatton, began exploring ways to use oblique sounders as a tool for determining optimum frequencies "on line."[56] Warren had been involved in these investigations when he was approached about the S-27 project. Rampage, carried out under Hatton, set out to demonstrate the possibility of choosing optimum frequencies by directly measuring propagation characteristics along a given path.[57]

For Hatton, the unreliability of short-wave stemmed from using ionograms to anticipate the often-dramatic variations in the high-latitude ionosphere.[58] His idea for an information/communication system sought to do away with the visual intermediary altogether. It was a vision of short-wave communications stripped to its bare necessities. Even the "conventional" radio sounding systems, which provided propagation information over the entire high-frequency spectrum, were too much. Hatton and his colleagues instead proposed what they called "channel sounding"—the focused evaluation of specific channels allocated to particular communication units. He drew on emerging information theory to propose pro-forma systems, carrying "go/no go" messages (necessary for nuclear response) that would reduce radio traffic by using prerecorded messages for routine information.[59] The main line of his research, though, focused on providing real-time propagation information for high-frequency radio operators, doing away with the need for prediction charts and the time-consuming and unreliable practices that accompanied them.[60]

In December 1960, Hatton put forward a system that combined sweep-frequency oblique sounding, channel sampling, and a communications system—a system very similar to the one Gerson would propose in Italy the following year. In it, the propagation path was sounded on all frequencies and the information was used to select the best frequency. The error rate was then measured on that channel. If the measurements were acceptable, communications began. The principles he helped lay out would inspire the American program to incorporate oblique data into real-time operation of circuits.[61]

Hatton's findings might have languished if it had not been for the interest of the Royal Canadian Air Force. In the late 1950s it had acquired a series of sophisticated medium and long-range aircraft to meet new operational requirements under NATO and the UN. A few were intended for bulk supply operations and troop transport, but a sub-set were directed toward the growing threat from Soviet nuclear submarines in the North Atlantic and the Arctic. US Naval Intelligence had already reported in 1952 that submarine-launched missiles would be the most likely means of nuclear attacks against the continental United States in a future war.[62] The warning motivated much of the sweeping program in oceanographic research that dominated the 1950s and was featured so prominently in the IGY. It also spurred the development of specialized anti-submarine aircraft to patrol the coastal waters of the continent. The most advanced of these were equipped with magnetic anomaly detectors, which, at low altitudes, could detect deformations in Earth's magnetic field caused by metal objects as small as a boat engine.[63]

The effectiveness of these aircraft, Air Force officials argued, was hampered by a characteristic problem—the frequent loss of radio contact between patrol aircraft and ground stations surrounding the North Atlantic. Even optimal frequencies for mid-Atlantic paths, whose communications were more stable, could increase from 6 to 30 megahertz over a two-hour period around sunrise.[64] Co-channel interference from a crowded radio spectrum and sunspot correlations suggested disruptions would only worsen through the 1960s.[65] Unless the Air Force wanted its new technological capability wasted, it needed a way to rapidly and reliably transmit tactical data to its maritime units. As one top secret Air Force memorandum noted, "it is clear that, until a rapid, reliable and flexible communications system is brought into use, these new RCAF [Royal Canadian Air Force] aircraft of great operational potential—the Argus, the Neptune, the Cosmopolitan, the Hercules and the Yukon—cannot be exploited to the full."[66] Air Force officials therefore pushed for programs they believed would "optimize" short-wave communications and allow them to reach their full potential. Among the techniques they suggested for this optimization were automatic sounding techniques and adaptive detection that would vary transmission rates depending on propagation conditions.[67] Aware of DRTE's research in ground-to-air communications, officials approached the Establishment in early 1960 to have members of the Rampage project investigate and advise.

George Jull was put in charge under the overall supervision of Lou Hatton. Jull had for years been interested in using oblique incidence sounders to increase the reliability of short-wave communications in auroral zone circuits, particularly during polar cap absorption events and geomagnetic storms.[68] He presented much of his research at the AGARD meeting on arctic communications in Greece in July 1963, where Gerson presented work on polar radio noise and antipodal communications.[69] Through the Rampage project, Jull's team conducted a series of trials in the North Atlantic and the Canadian North using modified oblique sounders during both disturbed and undisturbed conditions. After investigations and a preliminary report, DRTE officials recommended a prototype communication system based on direct sounding information and tailored for the Argus patrol aircraft.[70] Despite repeated lobbying from DRTE, schematics in hand, the proposal languished because of flagging interest from the Air Force and the Department of National Defence. Hatton therefore decided to push ahead regardless and to try to enlist the Air Force's interest. In November 1965, the block diagrams were turned over to the Electronics Laboratory for development.

The system that emerged from the Electronics Laboratory was known as CHEC (CHannel Evaluation and Call). "Its basic function," its builders explained, "is to help the radio operator achieve the full potential of his communication equipment at all times." A stepped-frequency receiver and transmitter were installed at a ground station. A second stepped-frequency receiver was installed in the aircraft. In the first two seconds after activation, the ground receiver measured, quantized, and encoded the background interference on one of its discrete frequencies, or "channels." During the third second, the ground transmitter sent this information to the airborne receiver, which decoded the interference information and measured its signal strength. Based on assumptions about propagation reciprocity, the receiver then "predicted" the signal-to-interference ratio that awaited airborne transmissions at the ground station. Through a bank of sixteen pairs of lights, the receiver then passed on the "essential" information to the airborne operator. If the evaluating transmission had been received successfully, the top light in a pair (corresponding to the assigned channel) would illuminate as an indication that simple transmission was possible. If the measured channel quality (the predicted signal-to-interference ratio) exceeded a lower limit set by the radio operator, the second light would illuminate, indicating that minimum communication conditions had been met. The radio operator could then proceed with the teletype transmissions of the message.[71]

The entire evaluation process for sixteen channels lasted 64 seconds and suggested a number of benefits. First of all, it avoided the time-consuming and often frustrating attempts to establish a communications channel by manual "hunt-and-try" techniques. The system was also a step toward disentangling the complex of human error, technical malfunction, and ionospheric variation that produced most communications failures. During an ionospheric blackout, for example, the system would simply register no open channels, and radio operators could be sure that co-channel interference, technical problems, or their own inexperience were not to blame. Similarly, if message reception was consistently lower than predicted, operators might look for likely faults in their equipment. Along with the need for an ionosonde, the requirement for special knowledge and abilities also fell away. As the system's engineers, Don Page and Bill Hindson, explained, "perhaps the most important potential advantage of the CHEC system as applied to air-ground communications is that the old requirement no longer exists for a veteran operator having long experience with the vagaries of the ionosphere."[72] For Page and Hindson, CHEC was above all a first step toward more comprehensive automation.

As it stood in 1967, CHEC was simply an aid to the airborne operator, who still had to select the best operating frequency from the display and then manually tune the transceiver to send the message. As a next step in the automation process, the engineers proposed that the system might automatically tune the transceiver to the optimal frequency. In an "ultimate stage" of automation, radio operators would punch their message on paper tape and feed it into the transceiver. Their task would then be complete. From here, the transceiver would take over, automatically reading the tape, selecting the optimal channel, and initiating the air-ground call. On the ground, the receiver would tune itself to the correct channel for reception, the ground station's transmitter would send acknowledgment, triggering the aircraft transmitter to send the stored message a given number of times depending on the predicted channel quality. "In this automated system," Page and Hindson concluded, "there is no longer need for a skilled radio operator."[73] In contrast to the efforts by early RPL personnel, the CHEC system required almost no knowledge of ionospheric propagation. It was the JANET system without the metaphysics of meteor trails.

The disappearance of special skill, though, shadowed the disappearance of the ionogram itself and its forms of visualizing the complex ionospheric phenomena of high latitudes. For Frank Davies, the greatest virtue of the new real-time, machine-based evaluation techniques lay in how "by the application of a special type of sounding mode to the transceiver, the need for an ionosonde is eliminated."[74] For two decades, the anxieties of Davies' group had been motivated precisely by the attempt to capture visually the complex phenomena of the high-latitude ionosphere as a way of under-standing and predicting its behavior. Faithfully representing the complex-ity of the ionosphere had been the goal; a lack of skilled workers capable of producing that representation on troublesome machines had been the impediment. The automatic ionosonde had been one solution to that com-bined problem. The new reading regimes had been a way of incorporating that complexity into the analytical machinery of the discipline. CHEC and its related systems sought to avoid representing that complex natural order altogether, finally realizing similar suggestions that had been made during World War II (discussed in chapter 2). E. E. Stevens, who as an employee of the Department of Transport's ionospheric station at Baker Lake had pro-duced the first panoramic ionograms taken from the Canadian Arctic, billed the evaluation tests as "a contest between two groups of communicators, the sounding-assisted team using CHEC and the radio officers using con-ventional methods."[75] As leader of the Arctic Predictions Section of the Communications Laboratory, Stevens had spent years poring over complex polar ionospheric data, detailing the difficulties of interpreting high-latitude records, and struggling with the inadequacy of derivative predic-tions and with the practical shortcomings of the prediction charts. Having helped to create the regime of machines and practices that made that complexity visible and analyzable, Stevens and his colleagues now sought to neatly and inconspicuously fold it into the technology of radio communications.

As the person responsible for the CHEC field trials, Stevens now believed that improving short-wave communications required stripping down the practice of radio operations so that the operator would be provided with only "the most essential information": signal-to-noise ratios and optimal transmission frequencies.[76] Ionograms and prediction charts, with their static and dated information, were both too simplistic (from the standpoint of ionospheric dynamics) and too complex (when viewed from the radio console). In place of the analog graphs, the CHEC display featured binary data and discrete choices for the operator that involved neither the interpretation of graphic forms nor the summoning of special expertise. Whereas researchers like Stevens had once sought to identify, classify, and understand radio disturbances through the higher-and-higher-resolution contours of the graph, they now hoped to conceal them in the binary low resolution of the double bank of lights.

Conclusion

In less than a year CHEC and its related projects were also canceled.[77] Even as Air Force officials authorized trials of the system in the spring of 1965, they doubted whether more effort should be put into a technology based on sounding and high-frequency improvements. In the end, officials in the Department of National Defence saw the technologies of short-wave radio and channel sounding as concepts whose days were numbered—outmoded and unreliable methods in an age when artificial communication satellites could make the ionosphere invisible while furnishing the reliability that the Cold War required.[78] Nathaniel Gerson had made similar arguments all along. For him, the adaptable real-time switching of CHEC stood alongside the very-low frequency systems designed for nuclear submarines, meteor-scatter communications of JANET, and massive cables running under the sea ice in an array of possible contenders for polar communications, each linked to its own set of prospects, limitations, and natural orders. Gerson ranked and evaluated them according to reliability and security. But he included a careful caveat that "all present long-range circuits may be replaced by satellite communication links if a sufficient number of communications satellites having polar orbits become available."[79] What tied together the various systems at play in polar communications in the early 1960s was how, through automation, they all involved the creation of technologies that worked in concert with the natural order, but in ways that increasingly obscured the ionospheric phenomena that the ionogram had made visible. Those systems remind us that there was a time when the concepts at the heart of survivable communications developed in explicit conversation with the natural orders of the Cold War—the lattice of meteor trails, the properties of ionospheric anomalies. But through automation they developed in ways that bypassed long-held practices of representing those orders, moving from the high-definition rendering of the ionograms to the low-resolution "displays" of teletype machines and banks of lights. Researchers continued to read in the effects and behavior of machines indications of the natural order on which they depended, but the large-scale associations of nature and technology at the heart of the "unreliable nation" were increasingly becoming a liability as the threats to the nation became internal.

7 Electromagnetic Geography and the Unreliable Nation

The efforts to map the ionosphere above Canada had always formed part of the broader state-sponsored attempts to map the postwar Canadian North, to make the region legible and governable, and to integrate it with the rest of Canada.[1] Those larger efforts marshaled the usual suspects of aerial photography, demographics, cartography, and natural resources behind them. But in the 1950s they also began to focus on radio signals as an explicitly geographical phenomenon. This complicated set of influences and interactions converged and was captured in the "radio geographies" of the period. In league with the projects to map the physical and human characteristics of the Canadian North, officials began to explore and chart the precise spatial extent of radio broadcast signals and how they related to natural phenomena, to human populations, and to the physical locations, behavior, and distribution of radio equipment throughout the region—the machinic orders of Northern radio communications.[2] The previous chapters of this book have charted the rise of the upper atmosphere as a mediating object between conceptions of hostile Northern nature and understandings of communications failures. This chapter retraces parts of that history, focusing on how the attempts to visually represent Northern radio not only interwove nature and machines in the ways that have been discussed so far, but also combined them with human geographies to define both the unreliability of radio transmissions and the Northern region itself.

Radio waves formed part of the built environment of the Canadian North. The radio navigation networks developed shortly after World War II had, of course, helped in the aerial mapping of the vast region.[3] But radio also represented what William Rankin has called an "intangible artifact"—part of a larger class of phenomena that includes odors, noise, and toxins. These artifacts are characterized by two common properties. First, they are temporally fleeting and thus rarely appear on maps, especially not on

topographic maps that show "permanent features." Second, they are difficult to contain. They are not just irreverent when it comes to national
boundaries, spilling across borders that halt other objects in their tracks.
As Rankin argues, they embody a profound misalignment between their
transient geography and more traditional political delineations of territory. Together, those two properties give radio waves an "insidious
power," allowing them to permanently occupy territories while appearing as temporary events.[4]

In view of that larger history, the radio geographies of the postwar North
are significant in a number of ways. Not only are they a rare historical
instance of mapping intangible artifacts against permanent topographical
features; but the state-sponsored act of mapping those artifacts revealed
anxieties about how their *impermanence* was disproportionately located—
contained in one sense—within geographic and topographical boundaries
that made it especially problematic. Northern radio signals formed a tempera-
mental infrastructure when laid over the more stable human geographies
clustered around river basins, estuaries, mining deposits, and natural ports
throughout the region. Their mapping was driven precisely by anxieties
about the manifest temporariness of those signals, the way their territorial
coverage only sporadically aligned with their target populations in the Canadian North. As phenomena, Northern radio signals made themselves felt, or
disappeared altogether, in inconvenient ways, at inopportune times, and at
a place and a historical moment where their uncertain temporal and geographic qualities could not be ignored. As artifacts, then, radio geographies
documented this impermanence and mobilized it behind the priorities of
the state. They suggest we should push Rankin's understanding further to
explore not only the unruly and insidious power of radio artifacts—how
they can be made invisible or tangible as needed—but also the politically
potent geographies that their impermanence has made possible.

The attempts to map the behavior of electromagnetic signals and the
performance of radio sets in the Canadian North bore important similarities to other cartographic traditions, particularly the medical geographies of
the nineteenth century. Those spatial understandings of health and disease
had enlisted physical geography, climatology, and meteorology to establish
complex relationships between illness and place, tying them, often explicitly, to Humboldtian physical geography with its isolines, geographically
dispersed precision measurements, synoptic studies, and mapping practices.[5] Like many other mapping efforts, medical geographies sought to
both document and transform. By making the distribution and dynamics

of diseases visible on maps, they sought to control, manage, and, ideally, eradicate them.[6]

Radio geographies shared these impulses toward documentation, correlation, and remediation. Like medical geographies, they combined phenomena from different registers—physical geography, atmospheric phenomena, human populations, and the distribution and behavior of machines.[7] Folding electromagnetic waves into broader geographic understandings of the region, they depicted those relations on official state maps and placed them (either visually or textually) alongside traditional concerns that included population, culture, health, economics, and settlement. Like those earlier geographies, too, they were motivated by concerns about race, authority, bureaucracy, and the relations of the state to the phenomena they depicted. Unlike those geographies, however, they eschewed causal speculation, focusing instead on correlations. Radio geographies were never a product of a single expert field or of a recognized profession, even though they were allied with state-sponsored geographical work and drew on state infrastructures, advanced cartographic techniques, and macro-scale mapping projects in the region. Instead, the radio geographies discussed here were *objects*—the products of a convergence of existing mapping practices, interventionist government ideologies, and the detailed measurements of electromagnetic phenomena throughout the Canadian North. They illustrated the troubling machinic order of Cold War Northern radio broadcasting as a way of amending it. As such, they helped define the unreliability of radio, the idea of North, and the perceived vulnerabilities of the nation that we have been exploring.

Beginning with the rise of the radio geographies, the chapter examines how the relationship between Northern nature and machine behaviors, elaborated through and around ionograms at DRTE, figured in the large-scale state communications initiatives of the same period. The research projects of DRTE developed alongside a changing human geography of the North that underwrote and intersected with it in complex ways. The chapter argues that, starting in the mid-1950s, at the height of RPL's efforts to define a distinctive natural order linked to radio disruptions, this changing human geography was used to define and give meaning to communication failures in the region. The concentration of particular groups of people, their relations to particular types of radio equipment, and their location in specific sub-regions of the North helped produce the unreliability of Northern radio. By the late 1950s, radio geographies defined the failure of radio broadcasts according to a human geography of the North, a spatial distribution of radio transmitters, and a natural order of high-Northern latitudes.

That fallibility, in turn, defined the North itself for the purposes of broadcasting. In the efforts of government officials and the Canadian Broadcasting Corporation, the "North" became, *by definition*, the region out of reach of regular and reliable broadcasts. That association between radio and the North extended beyond territory to the inhabitants of the region, as government officials increasingly envisioned radio as a medium "naturally" suited to the customs and culture of indigenous peoples. In the late 1960s, when satellites promised to overcome broadcasting disruptions, satellite communications were seen as a threat not only to the way of life of these peoples but also to the definition of the region itself.

The trajectory of the chapter, like that of the book, goes from threat to threat. It begins with the preoccupations that made Northern broadcasting a priority in the period—the threat of foreign indoctrination and cultural imperialism, the attempts to "open up" the North, and anxieties about national identity and sovereignty. It traces the preoccupations of government officials and broadcasting representatives through the creation of radio geographies that brought together natural phenomena, the distribution of machines, and the geographic concentrations of humans. It then turns to the CBC Northern Service, created to extend national broadcasting to the North, and explores how it used the unreliability of broadcasting to define the region and to create a proposed sympathy between the properties of radio and the needs of Northerners. The last section examines how the early anxieties about the integrity of the nation and the importance of reliable communications were transferred onto the emerging discord of the late 1960s and the early 1970s and reinterpreted as a question about the internal forces that threatened to destroy the country.[8] With the North still imperfectly connected to the nation, and with the rising threat of Quebec separatism, officials, including members of DRTE, turned to satellite communications as a solution to the unreliable nation. That shift signaled an end to the electromagnetic geographies of terrestrial radio and their definition of the North, and with it the end of DRTE.

Broadcasting and Territory

Broadcasting in the Canadian North started with a single wireless link established in 1923 between Dawson City and Mayo in the Yukon. It quickly grew during the interwar years into a regional communications network known as the Northwest Territories and Yukon Radio System. The system was intended to provide communications for the mining districts of the Yukon, the government of the Northwest Territories, and the fur trade.

The military was chosen to build it because no private commercial enterprise would shoulder the cost. To avoid the expense of land lines and maintain a body of trained radio teletype operators, the Royal Canadian Corps of Signals created a system of radio-telegraphs fanning out from the southern terminal at Edmonton. With the expansion of aviation and the discovery of gold and uranium in the 1930s, the system soon spread along the Mackenzie River to Aklavik, on the Arctic Ocean; and again during World War II as part of the infrastructure for the Northwest Staging Route and postwar preparations for continental defense, distant-early warning radar, military operations, weather stations, airfields, and winter exercises.[9] It incorporated Whitehorse and other booming population centers, but also remote locations such as Baker Lake, a Hudson's Bay Company trading post supplied once a year by ship, where the RPL ionospheric station would be located and where E. E. Stevens would produce the first "arctic" ionograms.

The system grew alongside a changing human geography that increasingly concerned government officials after World War II. By the late 1950s—amid a resource boom, a collapsing fur trade, and an expanding state that saw indigenous mobility as incompatible with state administration—a significant non-Native population occupied the towns and large administrative centers of the Yukon and Northwest Territories, a growing and increasingly permanent Native population settled around their outskirts, and seminomadic groups such as the Inuit of the Eastern Arctic and Subarctic moved in the vast spaces in between.[10] The resulting contrasts and tensions in this human geography—between indigenous and outsider, settled and itinerant, dominant and marginal—would profoundly shape the radio geographies of the region.

In the mid-1950s, government officials began to show interest in the geographical features of radio, with a view to integrating Northerners more fully with the rest of the country.[11] The Yukon and Northwest Territories and the "provincial norths" belonged nominally to their federal or provincial jurisdictions; but for most practical purposes they stood apart, incompletely integrated into the functions of the state. The inadequacy of Canadian broadcasts into the North from "outside," particularly from the Canadian Broadcasting Corporation, was used to signify the region's isolation from the rest of the nation. Minister of Northern Affairs and National Resources Jean Lesage, later a premier of Quebec and the architect of its Quiet Revolution, was shocked in 1954 when he toured the North and witnessed firsthand the almost complete absence of CBC broadcasts. Taped broadcasts of Canadian news came into the region days or even weeks after the events they reported. For Lesage, the asynchrony not only situated

Northerners out of time with the rest of the country; it also impaired radio education of Natives and Inuit located outside the catchment of local schools and the administrative grasp of the government. Just as troubling for Lesage, the absence seemed limited specifically to *Canadian* radio sources. He was deeply troubled by the prevalence of foreign short-wave broadcasts, a product of the escalating battles for short-wave dominance between the BBC foreign service, the Voice of America, and what Lesage saw as the "insidious and clever propaganda" of Radio Moscow.[12]

In the face of those concerns, the Canadian government lacked detailed geographic knowledge about how radio reception mapped onto human populations and territory. By 1950, DRTE had produced a number of general studies on reception conditions in the North, and a number of detailed reports on specific locations.[13] But these had mostly focused on military point-to-point communications. Accounts that made the broadcasting situation in the North immediately visible were scant. In their absence, government officials felt unsure about how to organize humans, territory, and electromagnetic waves effectively in producing not only people capable of being governed but people who were governed in fact.

The emerging radio geographies, like many other state-sponsored geographies, aimed to both document and transform. Lesage had appealed to the CBC to do something about the state of Northern radio. In the spring of 1955 the CBC revived a proposal to use the Northwest and Yukon system as repeater stations, beaming short-wave transmissions to them from a powerful transmitter somewhere on the southern edge of Western Canada, and then rebroadcasting locally. To study the possibility, the CBC sent two of its representatives—William Roxburgh from the Engineering Division in Montreal and Andrew Cowan, a former CBC war correspondent who had served in North Africa and Europe and was now in charge of Troop Broadcasts in Ottawa—on a three-week late-summer tour of Northern sites chosen because of their relation to the auroral zone and their possible problems with radio reception. Roxburgh's and Cowan's task was to make "on the spot" observations that included field measurements and local testimony. Their respective reports would mobilize human populations, the distribution and behavior of radio equipment, and a collection of natural phenomena to explain the fallibilities of Northern broadcasting.

Cowan had come to broadcasting after a varied career, having spent much of his youth in Alberta and rural Saskatchewan. He graduated from the University of Glasgow before joining the Brookings Institution (a Washington "think tank" located on Embassy Row and specializing in social sciences, particularly economics). After studying at Harvard University, he had taken a position as an economist with the British Royal Commission on the Highlands and Islands of Scotland, then a position on the secretariat of the International Wheat Advisory Committee in London, before organizing various CBC initiatives before World War II. After the hostilities, he coordinated the CBC's rehabilitation broadcasts for returning veterans, returning to London in 1946 as the CBC's European representative and then, in 1954, taking over Troop Broadcasts in Ottawa.

In his report, Cowan focused heavily on the way the radio stations were run, the amount of programming they received in the form of tapes or discs, and the quality and sources of Canadian broadcasts in particular, which were uniformly poor. But he wove the local conditions of broadcasting stations into the broader economic histories of the settlements they visited. Dawson City, the former capital of the Yukon and a center for the Gold Rush, had the air of a ghost town, he explained, where historic buildings such as the dance hall—"a frontier compromise between the old Bal Tabarin, and an 18th century German Court Theatre"—were being torn down before they crumbled.[14] Yellowknife, by contrast, had "an air of frenetic prosperity." The gold mining operations that supported it had swelled the population to four or five thousand; the town had spread from the rocky shores of Great Slave Lake to the overlooking plateau, and shaped the horizon with its smokestacks. But Cowan's often-careful accounts of the social geographies of the towns, including their populations, passed over their micro-geographies of inequality.[15]

In Roxburgh's analysis, these social, topographical, and economic elements became the canvas for a geography of radio devices. The engineer started with a detailed depiction of the spatial distribution and technical details of transmitters and receivers—their distance from the town proper; their proximity to roads and infrastructure; the arrangement of their antennas; their type and state of repair; the people in charge of their operation and maintenance. But he also imagined how those devices might be multiplied and extended, reinterpreting the topography of a town like Whitehorse in terms of an imagined future geography of radio. The town itself lay in a river valley 200 feet below the military bases and airport to the West, where a potential transmitter might be located; the three-story federal building, which housed the Territorial Administration, could provide a platform for transmission antennas. Roxburgh measured electromagnetic field intensities, invoked aurora and ground conductivity, and gathered reception reports from "responsible individuals" to construct a picture of radio reception in the North as a function of transmission source, time of day, season, and geographical position.[16] Canadian medium-wave broadcasts originating outside the North could not be picked up at all, he learned; equivalent American stations in the Pacific Northwest and Alaska and foreign short-wave stations were heard clearly, with reception improving from the west as one moved north. "Radio Moscow," Roxburgh noted, "was monitored consistently throughout the survey. . . . I would estimate that the Russian transmission in English directed toward our North runs a close second to the A.F.H.S. [US] service from the standpoint of reliability and signal strength." Canadian transmissions emerged here as small, ebbing circles of local reception, surrounded by vast areas of patchy and unreliable short-wave reception. Shifting again from actual to imagined geography, Roxburgh drew on DRTE investigations to suggest that a transmitter in the area of Vancouver would provide the best solution to the North's broadcast problems.[17]

Cowan and Roxburgh translated those complex spatial relations into a stark cartography that again mixed the actual and the potential. Its foundation was a map showing the entire Canadian North from 60° north latitude to the North Pole, itself constructed through synoptic laboratory analysis of aerial photographs. The "Canadian grid method" used to construct it had been adopted because of the way it satisfied the combined demands of practical requirements, Northern geography, and local institutional constraints in Canada; its use mirrored the relationship between laboratory, fieldwork, and photographs that characterized the practices surrounding the ionogram.[18]

Adding their own elements to the base map, Cowan and Roxburgh presented a map that divided the North into two principal electromagnetic regimes. (See figure 7.1.) The first showed the localized, medium-frequency coverage of the Northwest Territories and Yukon system. In the areas around the local stations, technicians had fanned out with field-strength meters, like the ones Jack Meek had carried aboard the RPL Mobile Observatory, measuring the strength of the stations' signals at surrounding points, plotting them on existing maps of the Dominion, and joining them to outline the station's electromagnetic footprint. According to the medium-wave maps, each station provided radio coverage within a limited range. The circles ebbed and waned with atmospheric conditions, and therefore with the time of day. They, and the machines that generated them, were concentrated in the densely populated river valleys of the Yukon and Mackenzie basins. With a range of about 30 kilometers during daylight and 100 kilometers during darkness, they took in the residents of the major towns, the Native populations concentrated (read: segregated) at their outskirts, and the workers in the heavy industries lying within their economic watershed—all told, about half of the Yukon's roughly 10,000 residents and two-thirds of the Northwest Territories' population of 16,000.[19] The uniform, outermost circles represented the area that could be taken in if Roxburgh's imagined geography were made real.

Outside these circles lived the "scattered," the "remote," and the "isolated"—the smaller rural communities and the nomadic. On the vast spaces of the Eastern Arctic and Subarctic, the map showed the area to be filled by high-frequency coverage for educational broadcasts. Their focus was the temporary settlements and federal schools, particularly in the great arc spanning northern Hudson's Bay and the Ungava Peninsula. (See the lower panel in figure 7.1.) Out of reach of the regional stations and relying on short-wave receivers, they had no way of receiving a consistent broadcast signal from any Canadian radio station.

The maps provided a spatial ordering of humans, territory, and radio waves designed to make the area legible to government officials. Read in this way, the report suggested that the unreliability of Canadian broadcasts was produced not by some transcendent natural order, but by the geographical convergence of a human geography of nomadic, semi-nomadic, and settled peoples who had flourished along the rich river valleys and ore deposits of the Yukon and the Mackenzie rivers on one hand, and on the other a vast electromagnetic "dead zone" generated by the location and power of foreign radio transmitters, the weakness and disrepair of local transmitters and antennas, the scarcity of short-wave receivers, and natural phenomena—atmospheric noise, ground conductivity, and aurora—that made Canadian broadcasts into those same areas weak, erratic, or altogether absent.

The radio geography was quickly mobilized and elaborated in 1955 by the commissioners of the two territories, Gordon Robertson of the Northwest Territories and F. H. Collins of the Yukon. In their lengthy meetings with Cowan and Roxburgh, both had discussed the detailed impediments to Northern broadcasting, as well as their plan to link the North to southern broadcasts through a high-power transmitter. The Commissioners now submitted their own evaluations to a Royal Commission on Broadcasting.[20] Although their emphases varied according to the local needs of the territories, they converged on a number of points, including concerns about national identity, sovereignty, and the external threats poised to fill the void created by the distribution of people, natural phenomena, and equipment in the region.

The commissioners' reports presented radio as a technology whose defining characteristics—orality, low cost, and geographical extension—made it uniquely suited to Northern society and to the project of unifying residents while connecting them to the rest of the nation. The CBC had understood

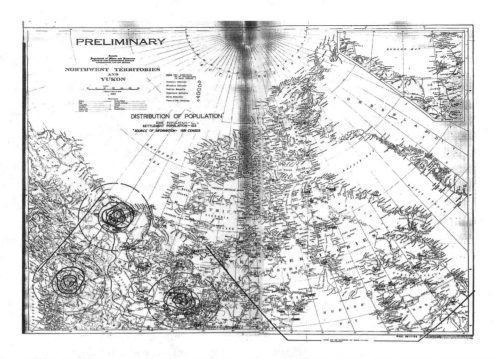

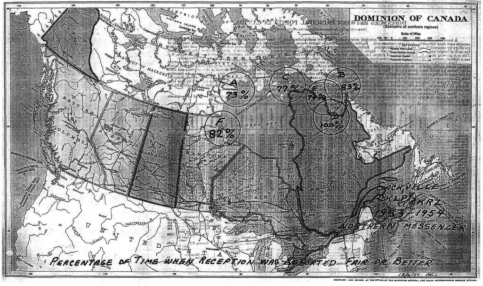

Figure 7.1

The electromagnetic geographies of Cowan and Roxburgh. The upper image shows the medium-wave coverage; the lower image shows short-wave reception in the area around Hudson's Bay. Sources: Cowan and Roxburgh, *Consideration of Broadcasting Coverage for the Yukon and Northwest Territories*, LAC, RG 41, volume 127, file 5–2; "Report On Reception of Northern Messenger Service from November 15th, 1953 to May 9th, 1954 inclusive," June 18, 1954, Library and Archives Canada, Record Group 41, volume 127, file 5–1. © Government of Canada. Reproduced with the permission of Library and Archives Canada (2016).

its target audience to be heterogeneous: a geographically dispersed popula-tion of 28,000 people in the two territories, most of them in the Yukon and Mackenzie Districts, with "a scattered two or three thousand Eskimos" in the Keewatin District of the Northwest Territories. In spite of this complex-ity, Cowan's and Roxburgh's report (which assimilated the desires of Native and Inuit groups to the opinions of town residents, government officials, and business leaders) had portrayed radio as an ideally uniform medium for that inhomogeneous population. Their view was echoed by Gordon Robert-son, who emphasized that radio programming in the rest of the Canada was as well suited to the North as to any other part of the country.[21] The desire of Northerners, Robertson asserted, was to become a more intimate part of Canada, and radio was the "most useful single method available" to achieve it. Robertson went on to explain that a dedicated service was needed for the North, "not because the needs of its listeners are any different from other Canadians, but because of the peculiar difficulties of transmission into the north and of broadcasting there."[22] His colleague, F. H. Collins, concerned almost wholly with the majority non-Native population of the Yukon and driven by assimilationism, cast the North in the same light. Although the situation would soon change dramatically, the radio geogra-phies of the late 1950s presented the people of the North as undifferentiated and undifferentiable when it came to radio broadcasts.

The reports also converged in their understanding that Northern broad-casting, in its current condition, embodied external threats to the nation and to the authority of the state. Robertson had served as Deputy Minister of Northern Affairs and National Resources and would rise to become the country's highest civil servant in the government of Lester B. Pearson. In the late 1950s he would put forward a vision for the Canadian Arctic as part of John Diefenbaker's "Road to Resources" program, expanding develop-ment schemes, social welfare programs, and defense projects across the North.[23] Robertson saw radio as a component in those developments, asserting that the cost of any broadcasting system would have to be borne by the Canadian state "as part of the cost of geography." No private enter-prise could afford the expense, and any national policy would have to extend comparable services across the country. A Rhodes Scholar and a vet-eran of the wartime Department of External Affairs, Robertson also hinted at darker forces that required the state's involvement. He described the broadcasts of Radio Moscow—transmitted every evening for about 30 min-utes and consisting of talk, music, and world news. "It is," he suggested ominously, "news obviously prepared by officials of the Soviet government with clear purposes in mind which need hardly be elaborated here." As one of the few sources of radio in the region, the Soviet broadcasts were particularly effective. "It would not be unreasonable to assume," Robertson continued, "that Radio Moscow may think that it has found its most exposed and therefore receptive audience in the Canadian north."[24]

The Yukon commissioner, F. H. Collins, reiterated his colleague's frustration about the way Northerners were isolated from the national life of Canada. In his meetings with Cowan and Roxburgh, he had been particularly concerned with "scattered and remote individuals and groups" that lay beyond the range of the transmitters at Whitehorse and Dawson. There were, he claimed, a disturbing number of people from those remote areas in mental hospitals. Referring readers to his map, Collins explained that the Yukon was roughly a triangle, bordered on two sides (from the point of view of radio) by the United States and on a third by a "radio vacuum." Soviet state broadcasts had begun to fill that vacuum, recognizing the region of 200,000 square miles as "an interesting battleground of Soviet and American ideologies through the medium of radio, while Canadian viewpoints are completely absent."[25]

The commissioners employed a cartography that made those threats overtly geographic. The year before their submissions to the Royal Commission on Broadcasting, as part of a study of the economic prospects of the territories, they had used standard maps created by the Geographical Branch of the Department of Mines and Technical Surveys. On them, technicians had depicted roads, fur trapping, and mining operations to explain the economic prospects for the territory. Those features were now gone. Overlaid on a geography that showed rivers, lakes, the size of local settlements, and the approximate limit of permanent polar ice were arrows—spectacular, in the case of the Yukon—entering the territory from the Soviet Union, the United States, and, as if it were a parallel foreign entity, Canada. (See figure 3.1.) It was unclear how accurately Collins or Robertson had sought to depict the strength and quality of the transmissions. But the maps followed military cartographic practices whereby the sizes of the arrows represented the strength of invading forces (in this case, short-wave broadcasts). The highly directed, invasive broadcasts of national powers contrasted sharply with the placid concentric circles of the local stations.

Probably working together, Robertson and Collins presented a common solution: a new high-power transmitter in the South. DRTE investigations had suggested that a high-power transmitter for Northern broadcasts would ideally be located near Vancouver, with a second one located near Winnipeg or Montreal if necessary.[26] The Vancouver facility would provide live connections to local stations in Whitehorse, Dawson, Watson Lake, Hay

River, Yellowknife, Norman Wells, and Aklavik. These local transmissions would also be heard, perhaps less regularly, on a great swath from Great Slave Lake to the Beaufort Sea, reaching about half of the territory's population. People outside those regions (particularly in areas around Hudson's Bay, including the Keewatin District, Baffin Island, and the Ungava Peninsula of northern Quebec) would be able to receive the broadcasts directly on short-wave sets.[27] A new Northern Broadcast Service would operate the short-wave stations and supervise the standard-band stations in the region, coordinating with the appropriate government departments.[28]

The Northern Service

The CBC Northern Service, created in 1958 and placed under Andrew Cowan's direction, was designed to solve the communications problems captured so strikingly in the radio geographies. Within its first six months, by May 1959, it had taken over stations at Whitehorse, Dawson, Yellowknife, and Goose Bay.[29] Its plans for a transmitter at Vancouver were derailed that summer when Canadian National Telegraphs announced that it would build land lines from Whitehorse to Mayo and from Hay River to Yellowknife, connecting a number of the existing stations and making a short-wave transmitter for the region redundant. Cowan responded by shifting the project eastward, reasoning that it would better serve the Eastern Arctic, which had no broadcasting service at all and whose population was "sparser, more widely scattered and more remote."[30] Cowan considered Winnipeg a likely site for a high-power transmitter, but it soon became clear that it might instead be possible to modify transmissions from the Sackville transmitter in New Brunswick, used for the broadcasts of the International Service. (See figure 7.2.) The 50-kilowatt transmitters had been installed in 1943 near where the waters of the Bay of Fundy fill the Cumberland Basin, creating the tidal Tantramar salt marshes. First French Acadians and then, after their expulsion from the United States, British settlers had reclaimed farming land from the marshes with a series of dikes and sluices. During World War II, the reclaimed land provided a nearly ideal site for receiving British broadcasts and for transmitting to Europe. The transmitters, finished in 1945, were connected to an enormous curtain antenna array of orange and white towers surrounding the station building to the north and west. (See figure 7.3.) They were linked to the CBC's studios in Montreal, 600 kilometers away, by land lines. After test transmissions in 1944, the short-wave service officially began transmitting in February 1945, with programs directed to Canadian troops overseas and European listeners. Tests

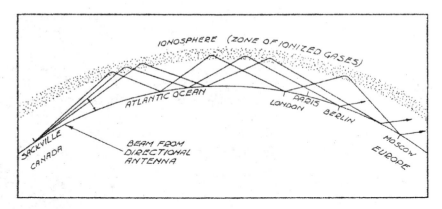

Figure 7.2

A diagram of transmission from the CBC transmitter at Sackville. The transmitter was originally used for International Service transmissions to Western and Eastern Europe, but was modified to provide broadcasting services to the North. Source: J. Alphonse Ouimet, "Canada's New Voice: Engineering Features of C.B.C.'s International Shortwave Station," *Engineering Journal* 28 (1945): 440–447. Reproduced with permission from University of Ontario Institute of Technology (2016).

conducted in 1953 and 1954 had already suggested that the reception in the eastern North was tolerable. Tests carried out again in 1959 indicated that, through antenna reversal, CBC engineers could improve reception across the vast area and into the Eastern Arctic.[31] The transmitter would direct two transmissions—one featuring mostly French broadcasts and directed toward Northern Quebec, Labrador, and Baffin Island, the other intended to cover the entire North and featuring programming in English, French, and Inuktitut. The CBC International Service would reschedule its evening transmissions to accommodate the new broadcasts.[32]

The shift to the powerful Sackville transmitters—used for propaganda broadcasts into *foreign* countries in a dozen languages, including Czech, Slovak, Ukrainian, and Russian—symbolized the North's imperfect integration into the rest of Canada. It also generated unexpected but predictable results. When the CBC International Service asked listeners in 1960 to report on reception quality in order to judge the quality of the new transmissions from the East Coast, it received the expected reports from DEW Line station operators and from residents of Baffin Island, Frobisher Bay, and Baker Lake; it also received letters from southern Ontario (not very far north), Pennsylvania, Nebraska, North Carolina, Puerto Rico, Cuba, Nicaragua,

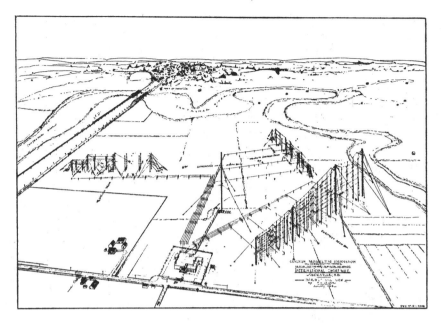

Figure 7.3
The layout of the Sackville site. The main transmitter was located in the building illustrated at bottom. Note the antenna arrays extending to the right. Source: J. Alphonse Ouimet, "Canada's New Voice: Engineering Features of C.B.C.'s International Shortwave Station," *Engineering Journal* 28 (1945): 440–447. Reproduced with permission from University of Ontario Institute of Technology (2016).

Colombia, Venezuela, Brazil, the Congo, and Japan. All the correspondents reported that the clarity of the signal was excellent.

In coming to terms with the region it served in the early 1960s, the Northern Service would also represent the culmination of the Northern radio geographies of the period. For two decades, archetypal Northern geophysical phenomena had been used to explain the unreliability of radio. DRTE had elaborated those connections in detail through the ionogram. Cowan, Roxburgh, and the territorial commissioners had joined many others in the late 1950s in repeating abridged versions of those explanations, linking atmospheric conditions to radio disruptions throughout the North. The Northern Service inverted that relationship, ultimately using the inconsistency of radio to define the North itself. In 1956, the CBC had defined "Northern Canada" as consisting of the area north of 60° latitude in Western Canada and north of Hudson's Bay and the Ungava peninsula in the East, stretching to the pole. By the end of the Northern Service's first year of operation, as Cowan's group acquired stations in Goose Bay, in Churchill, in Uranium City, in Cassiar, and in Fort Nelson, it had expanded that understanding to encompass most of the provincial norths: "North of the existing pattern of CBC network stations lies an area of nearly 2,000,000 square miles. It comprises the Yukon and Northwest Territories, and the northern part of all the provinces except the Maritimes. This great frontier is populated by an estimated 75,000 people, including 11,000 Eskimos and 6,000 Indians. In this area lie many of Canada's hopes for the future. Communications has always been one of its big problems." Cowan's group explicitly tied those communication problems to what contemporaries considered the immanent qualities of the North, particularly the geophysical phenomena being explored by DRTE. "Although radio would appear to be an easy answer to the problem of communications among the scattered population of the North," they explained in 1959, "this area is subject to some of the worst broadcasting conditions in the world. This is the result of atmospheric interference caused by the north magnetic pole, which lies within the territory."[33] By 1963, however, that unreliability—part of the electromagnetic geography laid out in the previous decade—had become constitutive of the region. The CBC now defined the Northern region as "extend[ing] north to the Pole and south to an imaginary line that would include those listeners *who do not receive a consistent and adequate broadcast signal* from CBC network stations or private stations located 'outside.' By this definition, it covers almost two million square miles and has, by the 1961 census, a population of about 80,000."[34] In 1965, as the network continued to expand and improve, Cowan dropped this definition and instead conceded that the territory "has no precise boundaries."[35] His attention now shifted to how radio might serve what he saw as the internal challenges facing the region and its people.

The Unreliable Nation

As the geographical focus of the Northern Service became less precise, Cowan began using radio to differentiate its inhabitants. His view of the natural sympathies between the North and radio would increasingly be transposed onto its indigenous peoples. That transfer signaled a bifurcation that was already underway. Until the mid-1950s, radio had been the only means of broadcasting in Northern Canada. With the rise of other possibilities in the mid-1960s, especially satellite communications and television, radio became explicitly linked to the life of indigenous peoples and was used to argue for a natural sympathy between them and the medium. That link would be woven into a larger anxiety about the instability of the nation itself as concerns about external threats to the nation were transferred onto the emerging social unrest of the late 1960s and the early 1970s and reinterpreted as worries about the internal forces threatening Canada (and other countries). Those threats were once again tied to questions of communications, and members of DRTE would play one last role in defining them.

Except for some planned expansion down the Mackenzie River, by the mid 1960s Cowan felt the technical problems of broadcasting into the North had been largely solved and that a plateau had been reached. His early focus on the expansion of broadcast facilities now shifted to what he saw as the needs of local people, and specifically to the role that radio could play for Native peoples and Inuit in the North. He set out to abandon the paternalistic and didactic approach of the past—"the traditional paternalism that marked relationships in the North for too long." But he held to the idea that the ultimate purpose of radio broadcasts was to elevate the economic position of native peoples while providing unity for the region in a time of drastic change.[36]

Cowan set out to identify social forms particularly well suited to the medium of radio that would help carry out that transformation. He looked specifically to the history of the CBC and found his sought-after models in its "listening groups." Started in 1941, these groups were initially the cornerstone of a project mounted within the Corporation by a southern Ontario farmer, Orville Shugg, who saw them as a way to turn young farmers into activists for rural issues. Groups of neighbors gathered weekly to listen to a half-hour program on a single farm issue and discuss the program using accompanying study guides, with their feedback providing material for subsequent shows. Launched as an educational experiment in a country that was still heavily rural, the groups were immensely popular and quickly formed the basis for postwar radio programs, including the National Labour Forum, which Cowan had helped organize. After a UNESCO study examined and documented the innovative Farm Forum techniques and published its findings in 1954, the format was adopted in India, in Ghana, and in France, but Cowan felt the format "was singularly well-suited to the Northern region." He furthermore felt that, as the most equitable of all the broadcast media, short-wave radio in particular provided an ideal medium: "It provides something for everybody. It serves the most isolated individuals and groups, including the bulk of the Eskimos, as well as the citizens of urban communities."[37] But Cowan went further, suggesting that radio possessed a natural connection to the condition, culture, and ways of life of native peoples. A "nucleus of potential listening groups," he noted, "exists in many Northern communities; Eskimo councils, Indian band councils . . . and (perhaps most importantly) co-operatives." By 1965, the publications of the Northern Service focused almost exclusively on these groups.[38]

Cowan elaborated his vision for Northern radio most fully in the spring of 1969, at a conference on Northern resources held in Whitehorse. His talk, "The Medium and the Message," borrowed its title from the Canadian communications theorist Marshall McLuhan. The theme running throughout Cowan's talk was that, for all its untapped natural resources, the North possessed an even greater wealth of undeveloped human resources. From the point of view of broadcasting, Cowan explained, the people of the North fell into two groups: "those who come in from 'outside' to develop its resources, administer it and minister to the physical and spiritual needs of the people; and those who were born here and most of whose ancestors lived here for countless generations."[39] The two groups—outsiders and natives—were like Disraeli's two nations in a single state, Cowan explained.[40] In Canada, they were separated by living standards, racial and cultural backgrounds, social and economic objectives, and, crucially, language. But for Cowan, they were also divided by their consonance with the technologies of television and radio, respectively. The first group was an economic and social elite drawn from Canada's majority culture. They had come to expect television, and government officials and companies eager to attract them to the North and keep them there exerted the strongest pressure on the CBC to extend television coverage to the region. The second group was largely made up of an economic underclass of Natives, Inuit, and Metis. Cowan saw radio as an ideal medium with which to reach them: "Radio is a ready made medium for a people whose cultural traditions are all oral, who have no written memory of the race of more than 2,500 years to draw on and have not been communicating with each other by the printed word for the past 600 years as we have."[41]

Cowan's attempt to both naturalize and essentialize the relationship between radio and Native forms of life traded on a deep fear of social unrest. The destabilizing threat from Soviet and American broadcasts that had dominated the submissions from the territorial commissioners was nowhere mentioned in his talk. Instead, Cowan posited, the threat now came from within. He hinted that the North, with its young and growing population and its dire economic circumstances, posed a great danger to the nation. Pointing to recent urban riots in the United States and quoting the recently murdered Martin Luther King Jr. on violence as the language of the inarticulate, he warned ominously that the turmoil and social upheaval engulfing American cities could spread to Canada, with the North as one of its flashpoints. Native peoples, he told his audience, needed a medium with which to discuss their situation among themselves, in their own languages if need be, and then on equal terms with the rest of the country. With rioting in American cities, Cowan saw radio as a universally accessible tool of peaceful self-expression and activism. Building on this vision of a technology of the indigenous, Cowan detailed his group's efforts to encourage Native peoples to adopt radio broadcasting and make it their own. If their programs were successful, Cowan concluded hopefully, they would contribute to the development of an articulate political power for the region.

Throughout his remarks, though, Cowan sensed that he was struggling against another vision. His initial arguments about radio's relationship to the North had drawn directly and indirectly on the work of DRTE. By the end of the 1960s, however, his views were increasingly at odds with one strain of thinking coming out of the Establishment. In 1966, John Chapman, the second-in-command at DRTE, had presented a very different relationship between emerging technology and the North. Chapman had steadily risen through DRTE since arriving at RPL as a summer student just after the war. During that time, he had championed a number of projects, including the work on "machine intelligence" that drove computer-aided analysis and the CHEC system. His crowning achievement, however, had been DRTE's scientific satellites. The technical aspects of the satellite program appealed to his interests in atmospheric investigations and engineering. But it was the satellite as a tool of politics more than science that increasingly drove his involvement.

Privately, Chapman had already seen spacecraft in this political role even before the launch of S-27. As early as August 1960, he had carefully argued that Canada should move into the field of communications satellites by the end of the decade. In 1962, as the Satellite Section was conducting its final tests on S-27, he helped author a DRTE proposal for "a purely Canadian navigation and communication system" using satellites in polar orbit. The proposal failed as a result of its expense, but on the recommendations of the resulting report Chapman negotiated the launch of four more scientific satellites—the Alouette-ISIS series—to build ties with NASA and additional engineering experience at DRTE.[42] In the late summer of 1966, with both Alouette I and II transmitting ionograms, and with Colin Franklin's mass-production system working overtime, Chapman addressed the General Assembly of the International Radio Scientific Union (URSI) in Munich, expressing disappointment that a "clear, readily-interpretable" picture of the ionosphere had still not emerged.[43] By then, however, he probably had abandoned the prospects of reliable ionospheric communications altogether.

In May, the Science Secretariat of the Privy Council asked Chapman to chair a comprehensive review of all upper atmosphere and space programs in Canada. The three-person committee included Peter Forsyth, who had completed his meteor-burst project with the Communications Laboratory and was now at University of Western Ontario. Their official review still invoked a distinctive natural order, advising that "the upper atmosphere over Canada is like that over no other country, with northern lights, ionospheric storms, and a magnetic field which trails away from the sun to nearly the moon's orbit." One of the main objectives of the emerging Canadian space program should therefore be to "study the unique atmosphere high over Canada." Explicitly, however, the document subordinated this goal to a prime objective for space technology: the "application to telecommunications and survey problems of a large sparsely settled country." Chapman went further, building a vision of the nation itself around the image of the communications and survey satellite: "Canada's struggle to maintain an independent identity while at the same time providing an acceptable standard of living for its people, leads us to develop competitive technologies, to train and educate scientists and engineers, and to resist firmly the erosion of national control over the essential fabric of our national structure." Cutting through the fuzzy prose were age-old concerns about the cultural and technical "risk of domination by the US," the same concerns that had supported HF research throughout the 1940s and the 1950s.[44] During the diplomatic debacles of the mid-1960s—disagreements with the United States about the Vietnam War in 1965 and about China's admission to the UN in 1967—Canadian officials sought political and economic independence from the United States. The gist of the committee's report was unequivocal: satellites should now be turned directly to the surveillance, surveying, and communications needs of the Canadian state.

Chapman would use the porous, hybrid, and volatile radio geography represented in Cowan's and Roxburgh's report and in DRTE's research to argue for a very different relationship of technology to the North. His study—delivered in Canada's centennial year, 1967—suggested how domestic satellite communications could realize the goal of reliable Northern communications. Against the incomplete patchwork of radio and television communications then in place, Chapman presented satellites as the answer to "truly national" broadcasting. They would make it practicable for television and radio programs to be received at any point in Canada as far north as 78°N, something that a UN collaboration with Sweden had shown was almost impossible with territorial broadcast systems such as short-wave.[45] Chapman singled out what he saw as the revolutionary effect on isolated communities—"the parts of Canada which have so far been without reliable communications"—particularly in the North.[46] He repeated Frank Davies' statements from twenty years earlier about the need for sovereignty over the Arctic archipelago and the other Northern territory. And he returned again and again to the extension of television. Chapman recognized that this medium, new to the North but commonplace in many other parts of the country, could be a major factor in accelerating the development of resources in isolated regions. Unlike Cowan, he drew no distinction between the populations of those areas, choosing instead to cast the satellite as a "unity tool" for the nation as a whole.

Positioning the satellite in this way was a calculated move. The year before Chapman began his study, the Royal Commission on Bilingualism and Biculturalism advised the Pearson government that Canada was unconsciously "passing through the greatest crisis in its history."[47] Together with working papers, reports, and correspondence, Chapman collected newspaper clippings on the rising separatism in Quebec that would inspire the creation of the Parti Quebecois in 1968 and culminate in the bombings, kidnappings, and killings of the October Crisis two years later. He and several of his DRTE colleagues spent time at Laval University in Quebec City learning French as part of a bicultural program for civil servants that had been launched by the federal government in 1967. The spring after Chapman's report was delivered, the Premier of Quebec, Daniel Johnson, had proposed to ally his province with the French-German Symphonie satellite project. One of his goals was to develop engineering expertise within the province, but the cultural justification was "to prevent the francophone population from being enclosed in a predominantly English system of communication."[48] Symphonie would allow programming from France to be beamed directly to Quebec, challenging the ability of the federal government to control satellite development exclusively. Chapman weighed in on the matter in September 1967, suggesting that the federal government might outmaneuver both France and Quebec by formalizing an agreement with Germany that would make it the sole authority with whom the German government could negotiate space matters. To maintain good relations with Germany, France would then be forced to honor the agreement instead of "causing mischief." Throughout his own report, as Chapman elaborated his own vision for satellite communications that might potentially cut across nations and continents, his views were framed by this sense of national crisis. The report paid particular attention to the role of satellites in providing nationwide French-language programming. Unlike Cowan, Chapman saw the North as only one element of a national project, and not even the most important one: "The provision of domestic satellite communications systems to provide both telecommunications to isolated areas and a truly national TV coverage in both French and English should be considered a major aim of the Federal Government."[49]

In an irony that only a bureaucracy could generate, Chapman's recommendations ultimately led to the dismantling of DRTE. As momentum gathered for non-proliferation of nuclear weapons and peaceful use of outer space, the Canadian government became eager to shed the appearance of military connections in its space activities, a motive that had also shaped the creation of NASA. It canceled support for long-standing research projects, including that of Gerald Bull, a brilliant but abrasive McGill engineering professor who had proposed using powerful cannons to launch satellites. Frustrated and resentful, Bull would eventually try to sell his inventions to Iraq before he was assassinated in March 1990 by British SAS operatives.[50] In April 1968, seizing on what it saw as the growing importance of communications in the midst of a project to modernize government agencies, the government of Pierre Trudeau moved to concentrate all its civilian space activities in a new Department of Communications.[51] To distance itself from the connections to ballistic missiles and defense communications, the government decided to effectively dissolve DRTE and transfer its research activities to the new department. Acting on Chapman's own recommendations, it authorized the reorientation of the satellite program and the development of a domestic satellite communications system for Canada.[52] The last of the Alouette-ISIS satellites, ISIS-C, was quickly canceled. In November, DRTE itself was transferred to the newly created Department of Communications (of which Chapman, incidentally, was named assistant deputy minister) as a first stage in its dissolution. The existing ISIS research satellites, DRTE's synoptic investigations, its electronics research, and its entire research program on the disturbed ionosphere were slated for transfer.[53] In the next few months, its projects were divided among the new Department of Communications, the Department of National Defence, and Canadian universities. The organization was officially dismantled on January 1, 1969.

That spring, just days after Andrew Cowan's talk in Whitehorse, Eric Kierans, the newly appointed Minister of Communications, acted on Chapman's recommendations and announced plans for the world's first geostationary communications satellite, called Anik ("brother" in Inuktitut, one of the languages of the Canadian Inuit). Kierans presented Anik as part of a new "Northern vision" for the coming decade. Its avowed aim was to provide television, telephone, and messaging to the North, and to bring those areas into the mainstream of the nation. Laurent Picard, vice-president of the CBC at the time, was assured that its main value would be to connect remote and Northern communities. But Kierans' own statements mentioned nothing of Cowan's anxiety about Native populations and the future of the country; he focused instead on how the device would confront separatism by creating the "world's first truly bi-lingual, multi-cultural society."[54] Recognizing the ambiguity, indigenous groups saw the satellite as an instrument of resource exploitation by a transient white population rather than as a tool for the region's historic inhabitants.[55] Invoking a developing domestic body of communications theory, academics joined in those concerns, explaining how the satellite would inundate Northerners (particularly Native peoples) with a southern culture and would jeopardize "more vital" types of communication such as radio.[56]

Those concerns about the way satellites jeopardized the character of the North played out in another register, reshaping once again the geographies of the region. Their most prominent stage was the work of the Quebecois geographer Louis-Edmond Hamelin. From his post at Laval University, as we saw in chapter 3, Hamelin had worked throughout the 1950s to develop the field of nordology, the study of the cold regions of the Northern Hemisphere. He had placed the concept of *nordicity*, "the state or quality of being north," at the core of the new discipline. In 1964 he quantified it with a "nordicity index" based on ten criteria that spanned the human-centered, the natural, and the geographic.[57] The index was based on latitude, permafrost, average temperatures, ice melt, vegetation cover, population (indigenous and white), exploitation of resources, cost of supplies, and "communications."[58] The poorer and less reliable a region's communications were, the more "truly Northern" it was. Throughout the 1960s and the early 1970s, as he directed Laval's Department of Geography, founded the Centre for Northern Studies, and served on the Council of the Northwest Territories, Hamelin patiently elaborated his index into a hemispheric system capable of measurement, comparison, and ranking.[59]

Hamelin reprised those definitional issues in 1975, now as an officer of the Order of Canada, when he returned to Laval and presented his geographical scheme in its most comprehensive form: a monograph, titled *Nordicité Canadienne*, that would win the Governor General's Award for nonfiction the same year. This work was not, Hamelin cautioned his readers, a geography in the normal sense. They would not find a steady succession of chapters devoted to climate, relief, ice, commerce, or natural environment; readers would search in vain for a systematic study of the Northern regions. The work was neither an atlas nor an encyclopedia.[60] Instead, it dealt with a series of problems posed by the concept of nordicity itself.

Hamelin's system began as an attempt to generate a useful and (in his view) more accurate understanding of what defined the North in Canada. But under the growing concerns about Native land claims and self-government in the early 1970s, he soon sought to mobilize that definition behind an overtly political project that defined the relationship between the North and the southern *écumène*, a term rich with Catholic and classical references not lost on Hamelin.[61] During his time in the Northwest Territories, Hamelin had commented on the tensions that informed Cowan's early human geography of the North. He spoke of a dislocation between "outsiders" and Native populations. "Fortunately," he ventured, "until now, the forces of tension have been mainly latent."[62] In the introduction to *Nordicité Canadienne*, he asked if the North was "to be assimilated into, integrated with, or even associated with the principal ecumene of the nation." Hamelin conceded that his masterwork was perhaps "too heavily devoted to the existential," but it was, at its core, a search for essences. By defining and quantifying nordicity, Hamelin sought to address a series of questions about authenticity: Was the Canadian North "Northern" in its political structures? Had the changed administration structures of the late 1960s involved a "true nordification"? Had the North in general been increasing or decreasing in nordicity over the last century?[63]

Hamelin revealed that modern trends tended toward "denordification." In the "Age of Discoveries," all of Canada had been considered a Northern country. Deploying his own index, he revealed how population growth, changing climate, industry, and infrastructure development had shrunk the boundaries of what was considered the Far North. Through "iso-nordic lines," he parsed the territory into gradations of nordicity that stretched from the Far North through the Middle North and finally to the thin populated strip of the southern ecumene, nearly to the latitude of Montreal. But central elements of his system had changed substantially by the mid-1970s. The concept of nordicity no longer focused on the state of "being north," but rather on "the state or level of being polar in the Northern Hemisphere." More tellingly for our story, the unreliability of "communications" no longer defined the region. By 1975—perhaps in recognition of the new role of broadcast satellites, which would have invalidated the category—"communications" had disappeared from Hamelin's scheme. They were considered "only indirectly," assimilated to a series of geographic, natural, and human factors.[64]

Conclusion

Throughout the twentieth century, radio waves reconfigured basic assumptions about geographic presence, occupation, and control.[65] Unlike the other symbolic technologies of the period—ICBMs, intercontinental bombers, and nuclear weapons—they could cross geopolitical boundaries without triggering the intricate mechanisms of collective defense and massive retaliation that characterized the period. Cold Warriors exploited this misalignment between the electromagnetic geography of the late twentieth century and the more conventional political geography of the Cold War, using it for propaganda, strategy, and spycraft.[66] But the incongruity also created enormous anxiety in Canada. The permanent occupation of Soviet and American radio broadcasts in the North reiterated the stakes of the Cold War and Canada's place within it. While government officials exploited the situation for their advantage, others worried about both the inability of Canada's own radio transmissions to occupy the North authoritatively and about the kinds of vulnerabilities that impermanence generated. The transient and faltering quality of Northern radio highlighted Canada's inability to create electromagnetic geographies that were coextensive with the territorial nation. Those technological anxieties around impermanence, instability, and failure generated their own entangled geographic understandings.

The radio maps of the 1950s captured one particularly powerful instance of these entangled geographies. Breaking with a cartographic tradition that separated radio waves from topographical features, they sought to document the state of Northern radio, weaving electromagnetic waves into the physical and human geography of the North. If Northern radio began as part of an attempt to exert territorial control over the postwar North, guarding it against physical incursions and occupations, it quickly became part of a dubious project to transform the Native populations of the North using the double-edged language of human potential. Not only did the radio geographies of the 1950s draw directly on DRTE's work; they paralleled its impulse to document the machinic order of Northern radio in order to overcome its failures. Although directed to different ends, each traded on understandings that underlay a broader conviction: that the qualities of the North inherently opposed reliable communications. For two decades, that opposition between the natural and the machinic had characterized DRTE's efforts to produce and interpret the ionogram. It had underwritten the Northern Service's definition of the Northern region, and had influenced Hamelin's attempts to quantify nordicity as a way of defining both the place of the North in Canada and Canada's place in the postwar world.

Those resonant projects had been mounted at a time when the Canadian North was the site of external threats—American challenges to sovereignty, the possibility of invasion by the Soviet Union, and the reality of electromagnetic incursions. As the perceived threats became internal—social unrest, native activism, and Quebec separatism—the discussion turned to which communication technologies could reliably ensure the survival of the nation. Even as Andrew Cowan awkwardly struggled to recast radio as a technology of the indigenous, relying on racial essentialism and fear to portray it as a solution to the potential social unrest in the region, the inherent links between radio and the North that he invoked again and again were being recast as political instruments for an opposing cause. The organization that had done so much to set up the inherent and defining tension between the North and communications, between the nation's ostensibly unique and hostile natural order and its problematic machinic orders, ultimately proposed to resolve it once and for all. Satellites were attractive precisely because they promised to realize the original DRTE project at a time when the nation again depended crucially on that realization. But in completing that project, the rise of satellites and the reliable communications they promised ultimately threatened the idea of North and the distinctive natural order that they had helped create. In time, opposition to satellite communications would fade as fresh forms of life grew up around the new machinic order, as the Sackville Towers were decommissioned and dismantled, and as "pure" electromagnetic geographies replaced the hybrids and patchworks of radio circuits that had characterized the region for half a century.

Conclusion

Throughout the modern period, nations defined themselves through the relationship between nature and machines. As abstractions, the natural and the technological were useful for thinking about difficult questions at the heart of the nation—questions about providence and contingency, inevitability and human agency, the historical and the transhistorical. They provided powerful, resonant, widely-understood vocabularies for describing historical forces beyond the scale of human action.[1] As concrete entities, on the other hand, nature and machines gave the abstract nation specificity and substance. The practices of science and natural philosophy naturalized nations, casting them as unique collections of natural objects and phenomena and making them appear both material and transcendent. Political pageants, ingenious inventions, and large-scale industrial infrastructures technologized nations, turning steam engines, nuclear reactors, and communications networks into embodiments of national will. But these two powerful ways of apprehending the modern nation—the natural and the technological—were anything but separate. They mapped onto and shaped each other throughout the modern period. The practices and ambitions that created nations as natural entities were repeatedly allied with attempts to understand them as technological spaces, and specifically as groupings of interconnected machines and their sometimes worrying behaviors.

Technological failures provided particularly resonant opportunities to work out those relations and their wider cultural meanings. Especially when they affected technologies of national interest, these faults, breakdowns, and malfunctions naturalized the larger-than-human dangers that defined nations, rallied their populations, and legitimized their actions. More than suspensions and interruptions, failures opened influential stories nations told about themselves.[2] Where many countries saw their history as a triumph of technology over the forces of climate, geography, and environment, others used technological failures to craft a powerful alternative identity, transforming failing machines into evidence of the distinctive local natures that caused them. Particularly at times when they worried about what defined them, modern nations portrayed themselves not just as a technological struggle against nature or a straightforward alliance with it, but as a systematically articulated group of relations—traced out in laboratories and field stations, in machine infrastructures and territorial maps—between dynamic national natures and the problematic behaviors of complex groups of machines.

Those historical processes are particularly important for understanding the Cold War. One of the profound legacies of World War II had been its detailed geographic understandings of how specific natures threatened the operations and survival of both humans and machines. The Cold War normalized that exceptional geography, expanding it to include the deep oceans and near space. It also made these hostile natures central to the logic of the conflict.[3] Although the most pressing threats of the period were military and geopolitical, they were constantly channeled through a language and geography of natural hostility and the dangers it posed to the period's most important technologies. These persistent, naturally-induced, and potentially apocalyptic weaknesses in global surveillance, communications, and nuclear response preserved the clarity of Cold War danger and helped sustain the conflict.[4] But at the national level, they helped create the Cold War's "states" of emergency: not just conditions of crisis, but also political formations that normalized imminent dangers and the role of the nation-state in confronting them.[5] Cold War strategists and technologists did not invent this use for hostile natures and technological failures. Ocean-going empires stretching back to the early modern period had used shipwrecks, weapons failures, and instrument malfunctions to define threatening "other" natures and to drive their political agendas.[6] But the Cold War amplified, honed, and expanded these historical legacies. It created machines of unprecedented complexity and of unrivalled potential for catastrophe.[7] And it used the real and potential failures of those technologies to define the natures that opposed it. The cases described in this book were particularly stark examples of a more general trend, repeated across modern geographies and classes of machines, that saw nature as technology made fallible.

In Canada, those relations took the form of a powerful antagonism between a hostile Northern natural order and the troublesome behavior of shortwave radio technologies throughout the country. Reacting to what they saw as the multiple threats facing the nation after World War II, government officials and defense scientists seized on the radio failures of the North—in all its territorial, cultural, and political ambiguity—as a way of leveraging geographical vulnerability into political power, influence, and distinctive identity. Within the project of a nation trying to reimagine itself in the postwar world, DRTE's research presented Canada as a natural order defined by distinctly "Northern" atmospheric phenomena and a technological space of uniquely powerful, widespread radio failures that threatened the technological integrity of the nation. While those associations drew on long-standing myths about the place of the North and of communications in the nation's history, DRTE's research inverted the traditional relationship between natural order, environment, and technology in Canada, taking the age-old sympathy between technologies and natural orders, recasting them as antagonists, and locating their point of opposition in the problematic behavior of machines that threatened the nation's survival. Although the project of reliable Northern radio failed as a communications initiative, it succeeded as way of repeatedly clarifying the stakes of the Cold War, characterizing its hostile natures, naturalizing its technological failures, and turning those failures to political advantage. As transient as those machine behaviors may appear, they had lasting effects: they signified not only external threats—from the Soviet Union and the United States—but also the fragility and vulnerability of an already improbable nation. In this way, this book illustrates one history, among many, of the culturally salient meanings technological failures acquire and the natural orders and national identities they help express.

These machinic orders have longer histories, though. As time-specific representations of particular machines and their specific properties and behavior, they come into being and pass away. Their temporariness reveals something deeper about the histories of science and technology and their relationship to the wider history of the modern period. Like natural disasters, machine failures are made to vanish too quickly from our modern thought.[8] Historically, the practices of science have contributed to that disappearance. Articulating the connections between nature and machines in Cold War Canada was not only about capturing distinctive phenomena on instrument displays and graphs; it was also about creating regimes of visibility that have been central to the history of science.[9] The early efforts of DRTE researchers involved laborious attempts to craft practices, devices, and ways of seeing that made high-latitude atmospheric phenomena both visible and valuable. Although those researchers used the Northern natural order to turn geographical vulnerability into political power, influence, and distinctive identity, their ultimate goal was to make those distinctive phenomena visible precisely in order to erase their effects on the behavior of machines. In this way, they formed part of a series of modernist cultural, scientific, and technical projects that aimed to document variation and difference as a means of eradicating them.[10]

The history of technology has often hinged precisely on this impulse toward invisibility. When machines function properly, they efface both natural and social conditions that make that functioning possible.[11] The Cold War raised the urgency of that erasure. Its most important strategic doctrines—first-strike capability, massive retaliation, mutually assured destruction—drove engineering philosophies that demanded hyper-reliability, creating machines capable of operating dependably across the world's hostile natures. During the Cold War, to help create these failure-proof machines, the earth sciences set out to document and register the most problematic environments and natural orders, hoping to support the creation of machines that would make entire classes of natural effects imperceptible.[12] That DRTE researchers cautiously embraced machine failures as a hallmark of their work, seeing failures as indicators of distinctive natural phenomena, making recalcitrant natures and failing machines central to their status and identity, speaks to the kinds of actionable spaces that the Cold War opened up at the intersection of individual choices, national histories, and global struggles. It also points to the broader place of technological failures in showing the historical connections between the regimes of visibility that have shaped the history of science and the regimes of invisibility that, in pursuit technical virtues including reliability, have guided the historical development of our most important technologies.

More generally, the episodes recounted here point us to the broader importance of technological failures, breakdowns, and malfunctions for history writ large. The modern period produced more than just a material culture in which machines became more complicated and dangerous; its broader transformations in work, probability, human nature, causation, and culpability shaped the way we explain and make sense of failing machines. One consequence of those transformations was that human causes, material causes, and natural causes became interchangeable in explaining why failures happened. For a single machine failure, Cold War engineers could interchangeably invoke careless humans, hostile natures, or faulty designs and materials.[13] In explaining disruptions in Northern radio circuits over the course of two decades, DRTE scientists singled out inexperienced personnel, degraded machines or Northern nature at different times and in different contexts. In view of this peculiar overdetermination of failures, the interesting *historical* question is not to discover why past failures *really* occurred, but rather to understand why, at specific times and places, particular constellations of natures, humans, or machines emerged as culprits. In this way, the history of failures and their meanings reveals the abiding concerns and cultural formations through which contemporaries made sense of their historical moments, limning cause and effect and tracing the boundaries of the natural, the social, and the technological. Far from arcane technical events or narrow forensic questions, these failures and their complex, sometimes contested, meanings belong to mainstream history.[14]

Many of the practices and projects described in this book were undoubtedly elements of a high modernism that is now coming to an end.[15] The impulse to map and predict atmospheric properties on a global scale, to monitor ionospheric disturbances and tie them to the functioning of an increasingly global collection of machines, was allied with now-unimaginable schemes—meteorological warfare, redirected ocean currents, artificially induced aurora—that epitomized the mid-twentieth-century assault on nature and the attempts to control it at all costs. But even as the modern nation loses its grip as a locus of identity and analysis, we can see those practices of association carried forward to continent- and world-spanning infrastructures—power grids, information networks, logistical systems—whose operation, disruption, and collapse continue to mix the natural and the machinic in ways that define them both.

Canada's place in those historical developments is not representative. The country's history and its geographical, political, and cultural status in the postwar world are too atypical. But its experience of technological failure highlights the complex tensions and motivations that surrounded the project of the Cold War and the ways in which hostile natures and failing machines could be used to craft identities, gain political advantage, and shape cultural meanings during the period. The late- twentieth century was filled with machine failures that pointed to these dangerous "other" natures—Soviet tanks lying sphinx-like in Afghan sands, early-warning radars blinded by Icelandic aurora, American machine guns muted by the dust and humidity of the Vietnamese jungle. Those artifacts and episodes provide us with occasions to "provincialize" the superpowers, talking about them only in reference to places and events normally understood as peripheral.[16] They invite us to write histories that otherwise would lie doubly obscured, lost to a modern memory that treats failures as objects that should be forgotten; and ignored by a posterity that treats their historical effects as fleeting. For those outside the centers of power, these enduring, all-too tangible failures have spoken volumes. And if history is any guide, they will speak again.

Notes

Introduction

1. For an examination of these questions, see Bernadette Bensaude-Vincent and William Newman, "Introduction: The Artificial and the Natural: State of the Problem," in *The Artificial and the Natural: an Evolving Polarity*, ed. Bernadette Bensaude-Vincent and William Newman (MIT Press, 2007).

2. See Benedict R. Anderson, *Imagined Communities: Reflections on the Origin and Spread of Nationalism* (Verso, 2006), chapter 4; David E. Nye, *America as Second Creation: Technology and Narratives of New Beginnings* (MIT Press, 2003); Carol E. Harrison and Ann Johnson, "Introduction: Science and National Identity," *Osiris* 24, no. 1 (2009): 1–14; Eugen Weber, *Peasants into Frenchmen: The Modernization of Rural France, 1870–1914* (Stanford University Press, 1976); Thomas P. Hughes, *Networks of Power: Electrification in Western Society, 1880–1930* (Johns Hopkins University Press, 1983); Chandra Mukerji, *Impossible Engineering: Technology and Territoriality on the Canal Du Midi* (Princeton University Press, 2015); Michael Adas, *Machines as the Measure of Men: Science, Technology, and Ideologies of Western Dominance* (Cornell University Press, 1990); Gabrielle Hecht, *The Radiance of France: Nuclear Power and National Identity After World War II* (MIT Press, 2009); Fa-ti Fan, "Nature and Nation in Chinese Political Thought: The National Essence Circle in Early-Twentieth-Century China," in *The Moral Authority of Nature*, ed. Lorraine Daston and Fernando Vidal (University of Chicago Press, 2004); Lisbet Koerner, *Linnaeus: Nature and Nation* (Harvard University Press, 1999); Judith Schueler, *Materialising Identity: The Co-Construction of the Gotthard Railway and Swiss National Identity* (Aksant, 2008); Grace Yen Shen, *Unearthing the Nation: Modern Geology and Nationalism in Republican China* (University of Chicago Press, 2014); Sara Pritchard, *Confluence: The Nature of Technology and the Remaking of the Rhône* (Harvard University Press, 2011); Johan Schot, Harry Lintsen, and Arie Rip, eds., *Technology and the Making of the Netherlands: The Age of Contested Modernization, 1890–1970* (MIT Press, 2010).

3. See, for example, Gregory K. Clancey, *Earthquake Nation: The Cultural Politics of Japanese Seismicity, 1868–1930* (University of California Press, 2006); Scott W.

Palmer, "On Wings of Courage: Public 'Air-Mindedness' and National Identity in Late Imperial Russia," *Russian Review* 54, no. 2 (1995): 209–226; David Biggs, "Breaking from the Colonial Mold: Water Engineering and the Failure of Nation-Building in the Plain of Reeds, Vietnam," *Technology and Culture* 49, no. 3 (2008): 599–623.

4. Fan, "Nature and Nation in Chinese Political Thought," 412.

5. Hecht, *The Radiance of France*.

6. For the more general use of nature to authorize the social, see Lorraine Daston and Fernando Vidal, eds., *The Moral Authority of Nature* (University of Chicago Press, 2004), 5; Raymond Williams, *Keywords: A Vocabulary of Culture and Society* (Oxford University Press, 1976), 219.

7. See Koerner, *Linnaeus*; Yen Shen, *Unearthing the Nation*; Karen Oslund, "'Nature in League with Man': Conceptualising and Transforming the Natural World in Eighteenth-Century Scandinavia," *Environment and History* 10, no. 3 (2004): 305–325.

8. See Hughes, *Networks of Power*; Hecht, *The Radiance of France*; David Blackbourn, *The Conquest of Nature: Water, Landscape, and the Making of Modern Germany* (Norton, 2006); Pritchard, *Confluence*.

9. For comparable attempts in other countries, see Paul N. Edwards, *The Closed World: Computers and the Politics of Discourse in Cold War America* (MIT Press, 1997); Jon Agar, *Science and Spectacle: The Work of Jodrell Bank in Post-War British Culture* (Harwood, 1998); Hecht, *The Radiance of France*. For other Canadian examples, see Tina Loo and Meg Stanley, "An Environmental History of Progress: Damming the Peace and Columbia Rivers," *Canadian Historical Review* 92, no. 3 (2011): 399–427; Daniel Macfarlane, *Negotiating a River: Canada, the US, and the Creation of the St. Lawrence Seaway* (University of British Columbia Press, 2014).

10. Matthew Farish, *The Contours of America's Cold War* (University of Minnesota Press, 2010); P. Whitney Lackenbauer and Matthew Farish, "The Cold War on Canadian Soil: Militarizing a Northern Environment," *Environmental History* 12, no. 4 (2007): 920–950.

11. See Robert Bothwell, *Alliance and Illusion: Canada and the World, 1945–1984* (University of British Columbia Press, 2008), chapter 3.

12. On the complexity of "nature," see Williams, *Keywords*; Donald Worster, "Doing Environmental History," in *The Ends of the Earth: Essays in Modern Environmental History*, ed. Donald Worster and Alfred W. Crosby (Cambridge University Press, 1988); William Cronon, "Modes of Prophecy and Production: Placing Nature in History," *Journal of American History* 76, no. 4 (1990): 1122–1131.

13. Peter Davidson, *The Idea of North* (Reaktion, 2005), 16–22; Also see Immanuel Kant, *Observations on the Feeling of the Beautiful and Sublime and Other Writings* (Cambridge University Press, 2011).

14. Davidson, *The Idea of North*, 51; Oslund, "'Nature in League with Man.'"

15. See, for example, Alexander von Humboldt, *Cosmos: A Sketch of a Physical Description of the Universe*, volume 1 (Longman, 1868), 13–14.

16. Patricia Fara, "Northern Possession: Laying Claim to the Aurora Borealis," *History Workshop Journal*, no. 42 (1996): 37–57; Oslund, "'Nature in League with Man.'"

17. The classic work here is Carl Berger, "The True North Strong and Free," in *Nationalism in Canada*, ed. Peter Russell (McGraw-Hill, 1966). Also see Carl Berger, "Canadian Nationalism," in *Interpreting Canada's Past*, volume 2: *Post-Confederation*, ed. J. M. Bumsted (Oxford University Press, 1993); Shelagh D. Grant, "Myths of the North in the Canadian Ethos," *Northern Review*, no. 3/4 (1989): 15–41; Renée Hulan, *Northern Experience and the Myths of Canadian Culture* (McGill-Queen's University Press, 2002); Sherrill Grace, *Canada and the Idea of North* (McGill-Queen's University Press, 2007); Janice Cavell, "The Second Frontier: The North in English-Canadian Historical Writing," *Canadian Historical Review* 83, no. 3 (2002): 364–389. On the *Pays d'en haut* of French-Canadian literature, see Davidson, *The Idea of North*, 191. On the expansion of science into the North as an extension of the state, see Morris Zaslow, *The Northward Expansion of Canada, 1914–1967* (McClelland & Stewart, 1988); William H. Katerberg, "A Northern Vision: Frontiers and the West in the Canadian and American Imagination," *American Review of Canadian Studies* 33, no. 4 (2003): 543–563. For a comprehensive history of scientific attempts to understand, exploit and protect the region, see Andrew Stuhl, Empires on Ice: Science, Nature, and the Making of the Arctic, PhD dissertation, University of Wisconsin, Madison, 2013.

18. In Canada, the "North" as a territorial and political formation is usually defined as the area lying above 60° north latitude. But the imaginative scope of that region has always been much larger. In this book, I use it (with the initial capital) to refer to the geographic region, the imaginative entity, or the natural order, depending on the context.

19. On "nature" as a term for describing environment, see Cronon, "Modes of Prophecy and Production." Cronon and many other environmental historians caution against equating nature too strictly with the nonhuman world; see Cronon, *Nature's Metropolis*. For prominent examples, see Donald Worster, "Transformations of the Earth: Toward an Agroecological Perspective in History," *Journal of American History* 76, no. 4 (1990): 1087–1106; Cronon, *Nature's Metropolis: Chicago and the Great West* (Norton, 1992); Cronon, "The Trouble with Wilderness; or, Getting Back to the Wrong Nature," in *Uncommon Ground: Rethinking the Human Place in Nature* (Norton, 1996); Richard White, *The Organic Machine* (Hill and Wang, 1996). For examples in history of technology, see James C. Williams, "Energy, Technology and the Environment," *Icon* 20, no. 1 (2014): 113–122; Sara Pritchard "An Envirotechnical Disaster: Nature, Technology, and Politics at Fukushima." *Environmental History* 17, no. 2 (2012): 219–243.

20. Northrop Frye, *The Bush Garden: Essays on the Canadian Imagination* (House of Anansi, 1995); Margaret Atwood, *Survival: A Thematic Guide to Canadian Literature* (McClelland & Stewart, 2004).

21. Frye, *The Bush Garden*, 220; Andrew Richter, *Avoiding Armageddon: Canadian Military Strategy and Nuclear Weapons, 1950–63* (University of British Columbia Press, 2002), 18.

22. For overviews, see Richard White, "American Environmental History: The Development of a New Historical Field," *Pacific Historical Review* 54, no. 3 (1985): 297–335; Williams, "Energy, Technology and the Environment." For a discussion of "envirotech," see Jeffrey K. Stine and Joel A. Tarr, "At the Intersection of Histories: Technology and the Environment," *Technology and Culture* 39, no. 4 (1998): 601–640; Hugh S. Gorman and Betsy Mendelsohn, "Where Does Nature End and Culture Begin? Converging Themes in the History of Technology and Environmental History," in *The Illusory Boundary: Environment and Technology in History*, ed. Martin Reuss and Stephen Cutcliffe (University of Virginia Press, 2010); Sara Pritchard, *Confluence* and "Joining Environmental History with Science and Technology Studies: Promises, Challenges and Contributions," in *New Natures: Joining Environmental History with Science and Technology Studies*, ed. Dolly Jorgensen and Finn Arne Jorgensen (University of Pittsburgh Press, 2013). For an alternative attempt to unite history of technology and environmental history, see Edmund Russell, James Allison, Thomas Finger, John K. Brown, Brian Balogh, and W. Bernard Carlson, "The Nature of Power: Synthesizing the History of Technology and Environmental History," *Technology and Culture* 52, no. 2 (2011): 246–259.

23. Northrop Frye, "From "Letters in Canada,"" in *The Bush Garden* (House of Anansi, 1995), 1.

24. Ibid. Also see Jody Berland, *North of Empire: Essays on the Cultural Technologies of Space* (Duke University Press, 2009); R. Douglas Francis, *The Technological Imperative in Canada: An Intellectual History* (University of British Columbia Press, 2009); Howard Fremeth, The Creation of Telesat: Canadian Communication Policy, Bell Canada, and the Role of Myth (1960 to 1974), MA thesis, Simon Fraser University, 2005; Marco L. Adria, *Technology and Nationalism* (McGill-Queen's University Press, 2010); Mary Vipond, *Listening In: The First Decade of Canadian Broadcasting, 1922–1932* (McGill-Queen's University Press, 1992).

25. See Lorraine Daston, "The Morality of Natural Orders: The Power of Medea," Tanner Lecture on Human Values, Harvard University, 2002. Daston distinguishes between local and universal natural orders. I treat the universal here as having particular local expressions useful to the nation. On the complex etymology of "nature," see Williams, *Keywords*, 219.

26. Lorraine Daston and Katharine Park, *Wonders and the Order of Nature, 1150–1750* (Zone Books, 1998), 2.

27. Lorraine Daston and Peter Galison, *Objectivity* (Zone, 2007), 19–20.

28. This view of technology as applied science is longstanding. See, for example, Mario Bunge, "Technology as Applied Science," *Technology and Culture* 7 (1966), 329–347; Ronald Kline, "Construing 'Technology' as 'Applied Science': Public Rhetoric of Scientists and Engineers in the United States, 1880–1945," *Isis* 86, no. 2 (1995): 194–122. For additional critiques, see Walter Vincenti, *What Engineers Know and How They Know It* (Johns Hopkins University Press, 1990); Ken Alder, *Engineering the Revolution: Arms and Enlightenment in France, 1763–1815* (Princeton University Press, 1997), 87.

29. For cautions against this declensionist narrative, see William Cronon, *Nature's Metropolis*; Tina Loo, "High Modernism, Conflict and the Nature of Change in Canada: A Look at 'Seeing Like a State,'" *Canadian Historical Review* 97 (2016): 34–58; Liza Piper, *The Industrial Transformation of Subarctic Canada* (University of British Columbia Press, 2010).

30. See Williams, *Keywords*; Donald Worster, *Nature's Economy: A History of Ecological Ideas* (Cambridge University Press, 1994) and *The Wealth of Nature: Environmental History and the Ecological Imagination* (Oxford University Press, 1994); Alfred W. Crosby, *Ecological Imperialism: The Biological Expansion of Europe, 900–1900* (Cambridge University Press, 2004); Edmund Russell, *Evolutionary History: Uniting History and Biology to Understand Life on Earth* (Cambridge University Press, 2011).

31. Suzanne Zeller, *Inventing Canada: Early Victorian Science and the Idea of a Transcontinental Nation* (McGill-Queen's University Press, 2009).

32. Daston and Vidal, *The Moral Authority of Nature*.

33. A future teacher to Marshall McLuhan, Innis would go on to elaborate an influential communications theory: a macro-historical view of political and economic control in which the character and extent of both "natural" and artificial communications technologies broadly conceived—media, railroads, waterways, trade routes— drove the consolidation and fate of states and empires. See Harold A. Innis, *Empire and Communications* (Dundurn, 2007) and *The Bias of Communication* (University of Toronto Press, 2008).

34. According to these connections, Innis reasoned, it was inevitable that Canada would remain British after the American colonies had initially broken away. Its economic marginality, its production of raw materials and its consumption of manufactured goods, furthermore determined its future, binding it first to the British and then to the American empires and setting the rough political boundaries of the nation. See Harold A. Innis, *The Fur Trade in Canada* (University of Toronto Press, 1999).

35. Ibid., 393. Innis' work anticipated historians' recent critiques of the opposition between natural environments and technology. On the general questioning of the antagonism between nature and technology, see Piper, *The Industrial Transformation*

of Subarctic Canada, 10–11. On technologies as interface between humans and nature, see James C. Williams, "Understanding the Place of Humans in Nature," in *The Illusory Boundary: Environment and Technology in History*, ed. Martin Reuss and Stephen H. Cutcliffe (University of Virginia Press, 2010); Donald Worster, "Transformations of the Earth: Toward An Agroecological Perspective in History," *Journal of American History* 76, no. 4 (1990): 1087–1106; Worster, "A Long Cold View of History: How Ice, Worms, and Dirt Made Us What We Are Today," *The American Scholar* 74, no. 2 (2005): 57–66; William Cronon, "Modes of Prophecy and Production"; Cronon, "A Place for Stories: Nature, History, and Narrative," *Journal of American History* 78, no. 4 (1992): 1347–1376; Ted Steinberg, "Down to Earth: Nature, Agency, and Power in History," *American Historical Review* 107, no. 3 (2002): 798–820; Linda Nash, "The Agency of Nature or the Nature of Agency?" *Environmental History* 10, no. 1 (2005): 67–69; Paul S. Sutter, "Nature's Agents or Agents of Empire?" *Isis* 98, no. 4 (2007): 724–754. For similar work in the history of technology, see Arthur F. McEvoy, "Working Environments: An Ecological Approach to Industrial Health and Safety," *Technology and Culture* 36, no. 2 (1995): S145; Pritchard, *Confluence*.

36. Jon Agar, *The Government Machine: A Revolutionary History of the Computer* (MIT Press, 2003), chapter 3.

37. On techno-nationalism, see David Edgerton, "The Contradictions of Techno-Nationalism and Techno-Globalism: A Historical Perspective," *New Global Studies* 1, no. 1 (2007): 1–32; S. Waqar H. Zaidi, "The Janus-Face of Techno-Nationalism: Barnes Willis and the 'Strength of England,'" *Technology and Culture* 49, no. 1 (2008): 62–88; Maurice Charland, "Technological Nationalism," *Canadian Journal of Political and Social Theory* 10 (1986): 196–220; Richard J. Samuels, *"Rich Nation, Strong Army": National Security and the Technological Transformation of Japan* (Cornell University Press, 1994); Patrick Joyce, *The Rule of Freedom: Liberalism and the Modern City* (Verso, 2003); James C. Scott, *Seeing Like a State: How Certain Schemes to Improve the Human Condition Have Failed* (Yale University Press, 1998).

38. Scott, *Seeing Like a State*; Agar, *The Government Machine*, 879.

39. On systems, see Hughes, *Networks of Power*; Agatha C. Hughes and Thomas P. Hughes, eds., *Systems, Experts, and Computers: The Systems Approach in Management and Engineering, World War II and After* (MIT Press, 2000); Thomas Hughes, "The Evolution of Large Technological Systems," in *The Science Studies Reader*, ed. Mario Biagioli (Routledge, 1999). For a critique of the systems approach, see John Law, "Technology and Heterogeneous Engineering: The Case of Portuguese Expansion," in *The Social Construction of Technological Systems: New Directions in the Sociology and History of Technology*, ed. Wiebe E. Bijker, Thomas P. Hughes, and Trevor Pinch (MIT Press, 1987).

40. For an alternative use of "technological order," see Robert M Hutchins, "Commentary," *Technology and Culture* 3, no. 4 (1962): 643–646; Gilles Deleuze and Félix Guattari, *A Thousand Plateaus: Capitalism and Schizophrenia* (University of Minnesota

Press, 1987). Here, like Deleuze and Guattari, I use the term to avoid a focus only on discrete machines. Unlike them, I do not use it to encompass the natural as an analogue of the organic.

41. On the sciences of the eye, see Lorraine Daston, "Objectivity and the Escape from Perspective," *Social Studies of Science* 22, no. 4 (1992): 597–618; Lorraine Daston and Peter Galison, *Objectivity*.

42. On the theoretical and experimental culture of ionospheric research, see Chen-Pang Yeang, *Probing the Sky with Radio Waves: From Wireless Technology to the Development of Atmospheric Science* (University of Chicago Press, 2015).

43. Hecht, *The Radiance of France*, 15. Also see Michael Thad Allen and Gabrielle Hecht, "Introduction: Authority, Political Machines, and Technology's History," in *Technologies of Power: Essays in Honor of Thomas P. Hughes and Agatha Chipley Hughes* ed. Michael Thad Allen and Gabrielle Hecht (MIT Press, 2001), 14; Paul N. Edwards and Gabrielle Hecht, "History and the Technopolitics of Identity: The Case of Apartheid South Africa," *Journal of Southern African Studies* 36, no. 3 (2010): 619–639. For an alternative usage, see Mitchell, *Rule of Experts*.

44. See, for example, John M. Staudenmaier, *Technology's Storytellers: Reweaving the Human Fabric* (MIT Press, 1985), 145–146. On the more general tendency of social theory to theorize connection and assembly, see Stephen Graham and Nigel Thrift, "Out of Order: Understanding Repair and Maintenance," *Theory, Culture & Society* 24, no. 3 (2007): 1–25.

45. See Lewis Mumford, *Technics and Civilization* (University of Chicago Press, 2010), 362. Mumford couched that observation in a telling insight: the same worldview that produced scientific objectivity was also responsible for "the relative passiveness of machine-trained populations" when faced with broken-down machines.

46. Bruno Latour, "Technology Is Society Made Durable," in *A Sociology of Monsters: Essays on Power, Technology, and Domination*, ed. John Law (Routledge, 1991). Also see the general introduction to *Shaping Technology/Building Society: Studies in Sociotechnical Change*, ed. Wiebe E. Bijker and John Law (MIT Press, 1992).

47. See Graeme Gooday, "Re-writing the Book of Blots: Critical Reflections on Histories of Technological Failure," *History and Technology* 14, no. 4 (1998): 265–291; Hans-Joachim Braun, "Introduction," *Social Studies of Science* 22, no. 2 (1992): 213–230; W. Patrick McCray, "What Makes a Failure? Designing a New National Telescope, 1975–1984," *Technology and Culture* 42, no. 2 (2001): 265–291; Kenneth Lipartito, "Picturephone and the Information Age: The Social Meaning of Failure," *Technology and Culture* 44, no. 1 (2003): 50–81; Elena Kochetkova, "A History of Failed Innovation: Continuous Cooking and the Soviet Pulp Industry, 1940s–1960s," *History and Technology* 31, no. 2 (2015): 108–132; Karel Davids, "Successful and Failed Transitions. A Comparison of Innovations in Windmill-Technology in Britain and the Netherlands in the Early Modern Period," *History and Technology* 14,

no. 3 (1998): 225–247; Philip Scranton, "Technology-Led Innovation: The Non-Linearity of US Jet Propulsion Development," *History and Technology* 22, no. 4 (2006): 337–367; Stuart W. Leslie and Robert H. Kargon, "Electronics and the Geography of Innovation in Post-war America," *History and Technology* 11, no. 2 (1994): 217–231; Barton C. Hacker, "The Gemini Paraglider: A Failure of Scheduled Innovation, 1961–64," *Social Studies of Science* 22, no. 2 (1992): 387–406. On failed projects and programs, see Maja Fjaestad, "Fast Breeder Reactors in Sweden: Vision and Reality," *Technology and Culture* 56, no. 1 (2015): 86–114; Helen Anne Curry, "Industrial Evolution: Mechanical and Biological Innovation at the General Electric Research Laboratory," *Technology and Culture* 54, no. 4 (2013): 746–781; Biggs, "Breaking from the Colonial Mold,"; Heather J. Hoag, "Transplanting the TVA? International Contributions to Postwar River Development in Tanzania," *Comparative Technology Transfer and Society* 4, no. 3 (2006): 247–267. On the contextual "failure" of technologies, see Erik M. Conway, "The Politics of Blind Landing," *Technology and Culture* 42, no. 1 (2001): 81–106. For an example of the literature on accidents (one consequence of failures), see Thomas R. Wellock, "Engineering Uncertainty and Bureaucratic Crisis at the Atomic Energy Commission, 1964–1973," *Technology and Culture* 53, no. 4 (2012): 846–884; Eda Kranakis, "Fixing the Blame: Organizational Culture and the Quebec Bridge Collapse," *Technology and Culture* 45, no. 3 (2004): 487–518; Gabrielle Hecht, "Enacting Cultural Identity: Risk and Ritual in the French Nuclear Workplace," *Journal of Contemporary History* 32, no. 4 (1997): 483–507; Charles Perrow, *Normal Accidents: Living with High-Risk Technologies* (Princeton University Press, 1999). On failure as a heuristic within engineering, see Henry Petroski, *Success through Failure: The Paradox of Design* (Princeton University Press, 2006); Petroski, *To Engineer Is Human: The Role of Failure in Successful Design* (Vintage Books, 1992).

48. Scott Sandage, *Born Losers: A History of Failure in America* (Harvard University Press, 2006).

49. See Staudenmaier, *Technology's Storytellers: Reweaving the Human Fabric*, 145–146; Lipartito, "Picturephone and the Information Age." On the constructed quality of these failures, see Steve Woolgar, "Configuring the User: The Case of Usability Trials," in *A Sociology of Monsters: Essays on Power, Technology, and Domination*, ed. John Law (Routledge, 1991); Gregory C. Kunkle, "Technology in the Seamless Web: 'Success' and 'Failure' in the History of the Electron Microscope," *Technology and Culture* 36, no. 1 (1995): 80–103; Ernst Homburg and Johna de Vlieger, "A Victory of Practice over Science: The Unsuccessful Modernisation of the Dutch White Lead Industry, 1780–1865)," *History and Technology* 13 (1996): 33–52; Gooday, "Re-writing the Book of Blots," 266; R. A. Buchanan, *The Power of the Machine: The Impact of Technology From 1700 to the Present Day* (Penguin, 1994).

50. Mumford himself would revisit the question of technological failure in this context in Lewis Mumford, *The Pentagon of Power: The Myth of the Machine*, volume II (Harcourt Brace Jovanovich, 1974), 411. Also see Perrow, *Normal Accidents*; David

Nye, *When the Lights Went Out: A History of Blackouts in America* (MIT Press, 2010); Simon Schaffer, "Easily Cracked: Scientific Instruments in States of Disrepair," *Isis* 102, no. 4 (2011): 706–717; David Edgerton, *The Shock of the Old: Technology and Global History Since 1900* (Oxford University Press, 2011); Roger Cooter and Bill Luckin, eds., *Accidents in History: Injuries, Fatalities, and Social Relations* (Rodopi, 1997); Paul Fyfe, *By Accident or Design: Writing the Victorian Metropolis* (Oxford University Press, 2015), 256; Jamie L. Bronstein, *Caught in the Machinery: Workplace Accidents and Injured Workers in Nineteenth-Century Britain* (Stanford University Press, 2008); Judith Green, *Risk and Misfortune: A Social Construction of Accidents* (UCL Press, 1997); John G. Burke, "Bursting Boilers and the Federal Power," *Technology and Culture* 7, no. 1 (1966): 1–23; Elaine Scarry, "Swissair 111, TWA 800, and Electromagnetic Interference," *New York Review of Books* 47, no. 14 (2000): 1–24; Diane Vaughan, *The* Challenger *Launch Decision: Risky Technology, Culture, and Deviance at NASA* (University of Chicago Press, 1996); Sara Pritchard, "An Envirotechnical Disaster: Nature, Technology, and Politics at Fukushima," *Environmental History* 17, no. 2 (2012): 219–243; Adriana Petryna, *Life Exposed: Biological Citizens after Chernobyl* (Princeton University Press, 2013); Peter Soppelsa, "Paris's 1900 Universal Exposition and the Politics of Urban Disaster," *French Historical Studies* 36 (2013): 271–298; Kim Fortun, *Advocacy After Bhopal: Environmentalism, Disaster, New Global Orders* (University of Chicago Press, 2001). On "compound disaster," see Tyson Vaughan, "Moving Beyond Social and Epistemological 'Bundan' in Fukushima," *Designing Media Ecology* 3 (2015): 52–58.

51. On the complex meanings of "working," see Wiebe E. Bijker and Trevor Pinch, "The Social Construction of Facts and Artifacts: Or How the Sociology of Science and the Sociology of Technology Might Benefit Each Other," in *The Social Construction of Technological Systems: New Directions in the Sociology and History of Technology*, ed. Wiebe E. Bijker, Thomas P. Hughes, and Trevor Pinch (MIT Press, 2012).

52. See, for example, Deborah R. Coen, *The Earthquake Observers: Disaster Science from Lisbon to Richter* (University of Chicago Press, 2012); Keith Wailoo, Karen M. O'Neill, Jeffrey Dowd, and Roland Anglin, eds., *Katrina's Imprint: Race and Vulnerability in America* (Rutgers University Press, 2010); Jürgen Buchenau and Lyman L. Johnson, eds., *Aftershocks: Earthquakes and Popular Politics in Latin America* (University of New Mexico Press, 2009); Charles F. Walker, *Shaky Colonialism: The 1746 Earthquake-Tsunami in Lima, Peru, and Its Long Aftermath* (Duke University Press, 2008); Fa-ti Fan, ""Collective Monitoring, Collective Defense": Science, Earthquakes, and Politics in Communist China," *Science in Context* 25, no. 1 (2012): 127–154; David Alexander, "An Interpretation of Disaster in Terms of Changes in Culture, Society and International Relation," in *What Is a Disaster? New Answers to Old Questions*, ed. Ronald W. Perry and E. L. Quarantelli (Xlibris, 2005); Geneviève Massard-Guilbaud, Harold Platt, and Dieter Schott, eds., *Cities and Catastrophes: Coping with Emergency in European History* (Peter M. Lang, 2002); Rebecca Solnit, *A Paradise Built in Hell: The*

Extraordinary Communities That Arise in Disasters (Viking, 2009); Theodore Steinberg, *Acts of God: The Unnatural History of Natural Disaster in America* (Oxford University Press, 2000); Eric Klinenberg, *Heat Wave: A Social Autopsy of Disaster in Chicago* (University of Chicago Press, 2015).

53. Susan Neiman, *Evil in Modern Thought: An Alternative History of Philosophy* (Princeton University Press, 2004), 240–250.

54. Coen, "Introduction: Witness to Disaster."

55. Gregg Mitman, "Where Ecology, Nature, and Politics Meet: Reclaiming the Death of Nature," *Isis* 97, no. 3 (2006): 496–504.

56. See the essays collected in *Katrina's Imprint*, ed. Wailoo et al., and in *Vulnerability in Technological Cultures: New Directions in Research and Governance*, ed. Anique Hommels, Jessica Mesman, and Wiebe E. Bijker (MIT Press, 2014). Also see Williams, "Understanding the Place of Humans in Nature"; Stéphane Castonguay, "The Production of Flood as Natural Catastrophe: Extreme Events and the Construction of Vulnerability in the Drainage Basin of the St. Francis River (Quebec), Mid-Nineteenth to Mid-Twentieth Century," *Environmental History* 12, no. 4 (2007): 820–844; William Cronon, "Introduction: In Search of Nature," in *Uncommon Ground: Rethinking the Human Place in Nature*, ed. William Cronon (Norton, 1996), 23–58.

57. On agency of nature, these references are too numerous to list. For a few early works, see William H. McNeill, *Plagues and Peoples* (Anchor Books, 1976); Donald Worster, *Nature's Economy*; Timothy Mitchell, *The Rule of Experts: Egypt, Techno-Politics, Modernity* (University of California Press, 2002); Alfred W. Crosby, *Ecological Imperialism*. More recently, historians of empire and science have taken up these themes; see, e.g., Richard Grove, *Green Imperialism: Colonial Expansion, Tropical Island Edens, and the Origins of Environmentalism, 1600–1860* (Cambridge University Press, 1995). On the agency of nature against the human constructs, see chapters in *The Nature of Cities* (University of Rochester Press, 2006).

58. See, for example, See Matthew Farish, "The Lab and the Land: Overcoming the Arctic in Cold War Alaska." *Isis* 104, no. 1 (2013): 1–29; Lackenbauer and Farish, "The Cold War on Canadian Soil."

59. See John Robert McNeill and Corinna Unger, eds., *Environmental Histories of the Cold War* (Cambridge University Press, 2013).

60. Richter, *Avoiding Armageddon*, 8–9.

61. See J. L. Granatstein, *Canada, 1957–1967: The Years of Uncertainty and Innovation* (McClelland & Stewart, 1986); Bothwell, *Alliance and Illusion*.

62. Richter, *Avoiding Armageddon*, 8–9.

63. George Orwell, "You and the Atomic Bomb," *Tribune* (London), October 19, 1945.

64. Odd Arne Westad, "Exploring the Histories of the Cold War: A Pluralist Approach," in *Uncertain Empire: American History and the Idea of the Cold War*, ed. Joel Isaac and Duncan Bell (Oxford University Press, 2012), 54. For a survey of this transnational history, see *The Cambridge History of the Cold War*, volume 1 (Cambridge University Press, 2010). Also see Gabrielle Hecht, ed., *Entangled Geographies: Empire and Technopolitics in the Global Cold War* (MIT Press, 2011), 6; Hunter Heyck and David Kaiser, "Introduction," *Isis* 101, no. 2 (2010): 363; Westad, "Exploring the Histories of the Cold War," 54.

65. The literature on Cold War science is vast. For some representative works, see Kim Dong-Won and Stuart W. Leslie, "Winning Markets or Winning Nobel Prizes? KAIST and the Challenges of Late Industrialization," *Osiris* 13 (1998): 154–185; Hiroshi Ichikawa, "Introduction: A Perspective on the Historical Study of Science and Technology during the Second World War and the Cold War in Japan," *Historia Scientiarum* 16 (2006): 1–4; Benjamin A. Elman, "New Directions in the History of Modern Science in China," *Isis* 98, no. 3 (2007): 517–523; Zuoyue Wang, "Science and the State in Modern China," *Isis* 98, no. 3 (2007): 558–570; Suzanne Moon, "Justice, Geography, and Steel: Technology and National Identity in Indonesian Industrialization," *Osiris* 24, no. 1 (2009): 253–277. Other exemplars of international and comparative perspectives include John Krige and Kai-Henrik Barth, "Introduction: Science, Technology, and International Affairs," *Osiris* 21, no. 1 (2006): 1–21; Gabrielle Hecht and Paul N. Edwards, *The Technopolitics of Cold War: Toward a Transregional Perspective* (American Historical Association, 2007); Richard Beyler, "The Demon of Technology, Mass Society, and Atomic Physics in West Germany, 1945–1957," *History and Technology* 19 (2003): 229–241; Alexei Kojevnikov, "The Phenomenon of Soviet Science," *Osiris* 23, no. 1 (2008): 115–135; Konstantin Ivanov, "Science After Stalin: Forging a New Image of Soviet Science," *Science in Context* 15 (2002): 317–338; John Krige, *American Hegemony and the Postwar Reconstruction of Science in Europe* (MIT Press, 2008); Jacob Darwin Hamblin, "Exorcising Ghosts in the Age of Automation: United Nations Experts and Atoms for Peace," *Technology and Culture* 47, no. 4 (2006): 734–756. On biological and environmental sciences, see Angela Creager, "Nuclear Energy in the Service of Biomedicine: The US Atomic Energy Commissions Radioisotope Program, 1946–1950," *Journal of the History of Biology* 39, no. 4 (2006): 649–684; Evelyn Fox Keller, *Making Sense of Life: Explaining Biological Development with Models, Metaphors, and Machines* (Harvard University Press, 2002); David Serlin, *Replaceable You: Engineering the Body in Postwar America* (University of Chicago Press, 2004).

66. For other studies of research sponsored by the Canadian government during the period, see, for example, Bruce Doern and Jeffrey Kinder, *Strategic Science in the Public Interest: Canada's Government Laboratories and Science-Based Agencies* (Uni-

versity of Toronto Press, 2007); Andrew B. Godefroy, *Defence and Discovery: Canada's Military Space Program, 1945–74* (University of British Columbia Press, 2012); Robert Bothwell, *Nucleus: The History of Atomic Energy of Canada Limited* (University of Toronto Press, 1988); Jonathan Turner, The Defence Research Board of Canada, 1947 to 1977, PhD dissertation, University of Toronto, 2012.

67. For a similar focus, see the essays in Hecht and Edwards, *The Technopolitics of Cold War.*

68. See, for instance, recent scholarship on the Cold War human sciences: S. M. Amadae, *Rationalizing Capitalist Democracy: The Cold War Origins of Rational Choice Liberalism* (University of Chicago Press, 2003); Hunter Crowther-Heyck, *Herbert A. Simon: The Bounds of Reason in Modern America* (Johns Hopkins University Press, 2005); Jennifer S Light, *From Warfare to Welfare: Defense Intellectuals and Urban Problems in Cold War America* (Johns Hopkins University Press, 2003); Jamie Cohen-Cole, "The Creative American: Cold War Salons, Social Science, and the Cure for Modern Society," *Isis* 100, no. 2 (2009): 219–262; Mark Solovey, "Project Camelot and the 1960s Epistemological Revolution: Rethinking the Politics–Patronage–Social Science Nexus," *Social Studies of Science* 31, no. 2 (2001): 171–206; Mark Solovey and Hamilton Cravens, eds., *Cold War Social Science: Knowledge Production, Liberal Democracy, and Human Nature* (Palgrave Macmillan, 2012).

69. On threat as a way of preserving the clarity of the Cold War, see Holger Nehring, "What Was the Cold War?" *English Historical Review* 127, no. 527 (2012): 920–949. On the metaphorical use of the Cold War, see G. Johnston, "Revisiting the Cultural Cold War," *Social History* 35 (2010): 290–307; Hugh Wilford, *The Mighty Wurlitzer: How the CIA Played America* (Harvard University Press, 2008); Helen Laville and Hugh Wilford, *The US Government, Citizen Groups and the Cold War: The State-Private Network* (Routledge, 2012).

70. See Hecht, *Entangled Geographies*. On technology as a site of Canadian accommodation and resistance, see Robert Teigrob, *Warming Up to the Cold War: Canada and the United States' Coalition of the Willing, from Hiroshima to Korea* (University of Toronto Press, 2009); Richter, *Avoiding Armageddon, 1950–63*. For the broader range of international responses, see M. P. Leffler, "Bringing It Together: The Parts and the Whole," in *Reviewing the Cold War: Approaches, Interpretations, and Theory*, ed. Odd Arne Westad (Routledge, 2000).

71. Northrop Frye, "Sharing the Continent," in *Divisions on a Ground: Essays on Canadian Culture*, ed. James Polk (House of Anansi, 1982), 57.

Chapter 1

1. I use "nature" here in the sense of natural order rather than environment. On the relationship between war and environment, see Laura A. Bruno, "The Bequest of the

Nuclear Battlefield: Science, Nature, and the Atom during the First Decade of the Cold War," *Historical Studies in the Physical and Biological Sciences* 33, no. 2 (2003): 237–260; Ron Doel, "Constituting the Postwar Earth Sciences: The Military's Influence on the Environmental Sciences in the USA After 1945," *Social Studies of Science* 33, no. 5 (2003): 635–666; Richard P. Tucker and Edmund Russell, *Natural Enemy, Natural Ally: Toward An Environmental History of Warfare* (Oregon State University Press, 2004); John R McNeill, "Woods and Warfare in World History," *Environmental History* 9, no. 3 (2004): 388–410; Paul N. Edwards, "Meteorology as Infrastructural Globalism," *Osiris* 21 (2006): 229–250; P. Whitney Lackenbauer and Matthew Farish, "The Cold War on Canadian Soil: Militarizing a Northern Environment," *Environmental History* 12, no. 4 (2007): 920–950; Charles E. Closmann, *War and the Environment: Military Destruction in the Modern Age* (Texas A&M University Press, 2009); Janet Martin-Nielsen, "The Other Cold War: The United States and Greenland's Ice Sheet Environment, 1948–1966," *Journal of Historical Geography* 38, no. 1 (2012): 69–80; John Robert McNeill and Corinna R. Unger, eds., *Environmental Histories of the Cold War* (Cambridge University Press, 2013).

2. Peter Galison, "Feynman's War: Modelling Weapons, Modelling Nature," *Studies in History and Philosophy of Modern Physics* 29, no. 3 (1998): 391–434.

3. Meeting of NDRC Division 5, March 30, 1944, 94–95, Richmond files, 62, quoted in James H. Capshew, "Engineering Behavior: Project Pigeon, World War II, and the Conditioning of B. F. Skinner," *Technology and Culture* 34, no. 4 (1993): 835–857.

4. Norbert Wiener, *I Am a Mathematician* (MIT Press, 1964), 231–252; Peter Galison and Bruce Hevly, eds., *Big Science: The Growth of Large-Scale Research* (Stanford University Press, 1994), 240.

5. The inverse is also true; weapons scientists have understood the workings of machines through biological metaphors. On the use of birth metaphors in nuclear weapons engineering, see Hugh Gusterson, "Nuclear Weapons Testing: Scientific Experiment as Political Ritual," in *Naked Science: Anthropological Inquiry Into Boundaries, Power and Knowledge*, ed. Laura Näder (Routledge, 1996).

6. For the longer history of radio technologies as weapons, see Daniel R. Headrick, *The Invisible Weapon: Telecommunications and International Politics, 1851–1945* (Oxford University Press, 1991).

7. Short-wave (or high-frequency) radio uses electromagnetic waves in the range of 3–30 megahertz.

8. On the Manichean Sciences, see Peter Galison, "The Ontology of the Enemy: Norbert Wiener and the Cybernetic Vision," *Critical Inquiry* 21, no. 1 (1994): 228–266. On the later use of behavioral science in characterizing the Cold War enemy, see Ron Theodore Robin, *The Making of the Cold War: Enemy Culture and Politics in the Military-Intellectual Complex* (Princeton University Press, 2001).

9. See Aitor Anduaga, "The Realist Interpretation of the Atmosphere," *Studies in History and Philosophy of Science B* 39, no. 3 (2008): 465–510; Chen-Pang Yeang, "From Mechanical Objectivity to Instrumentalizing Theory: Inventing Radio Ionospheric Sounders," *Historical Studies in the Natural Sciences* 42, no. 3 (2012): 190–234.

10. See Aitor Anduaga Egaña, *Wireless and Empire: Geopolitics, Radio Industry, and Ionosphere in the British Empire, 1918–1939* (Oxford University Press, 2009); Chen-Pang Yeang, *Probing the Sky with Radio Waves: From Wireless Technology to the Development of Atmospheric Science* (University of Chicago Press, 2015).

11. See Henry Guerlac, *Radar in World War II* (American Institute of Physics, 1987); Robert Buderi, *The Invention That Changed the World: How a Small Group of Radar Pioneers Won the Second World War and Launched a Technological Revolution* (Simon & Schuster, 1997); Louis Brown, *A Radar History of World War II: Technical and Military Imperatives* (Institute of Physics Publishing, 1999), chapter 2.

12. For comparisons between American and British ionosonde development, see Yeang, "From Mechanical Objectivity to Instrumentalizing Theory."

13. On the history of graphical methods more generally in science, see Robert Brain, "Graphical Method," in *Reader's Guide to the History of Science*, ed. Arne Hessenbruch (Routledge, 2013).

14. Martin J. Rudwick, "The Emergence of a Visual Language for Geological Science 1760–1840," *History of Science* 14 (1976): 149–195.

15. Raymond Heising, quoted in Anduaga, "The Realist Interpretation of the Atmosphere," 488.

16. On modularity in the work of the Rad Lab, see Galison, "Feynman's War."

17. R. W. P. King, H. R. Mimno, and A. H. Wing, *Transmission Lines Antennas and Wave Guides* (McGraw-Hill, 1945), 314.

18. Transcript of Canadian Radio Wave Propagation Committee Conference, October 22–27, 1945, 27, Library and Archives Canada, Record Group 24, volume 3412. (In subsequent notes, Library and Archives Canada is abbreviated to LAC and Record Group to RG.)

19. See Catherine E. Allan, "A Minute Bletchley Park: Building a Canadian Naval Operational Intelligence Centre, 1939–1943," in *A Nation's Navy: In Quest of Canadian Naval Identity*, ed. Michael L Hadley, Robert N. Huebert, and F. W. Crickard (McGill-Queen's University Press, 1996).

20. J. H. Dellinger and Newbern Smith, "Developments in Radio Sky-Wave Propagation Research and Applications during the War," *Proceedings of the IRE* 36, no. 2 (1948): 258–266.

21. On the longer history of similar global infrastructures in meteorology, see Paul N. Edwards, "Meteorology as Infrastructural Globalism," *Osiris* 21 (2006): 229–250.

22. M. M. Gordon, "Weekly Radio Forecasts," Confidential Report, October 26, 1944; unsigned secret letter to the Treasury Board of Canada, February 19, 1945, 1, LAC, RG 24, volume 3414, file 468-4-2.

23. F. T. Davies and J. H. Meek, "Report to CRWPC on International Radio Wave Propagation Conference, Washington, DC—April 17–May 5, 1944," presented May 11, 1944, 5, LAC, RG 24, volume 3413, file 468-3-2, volume 1.

24. Combined Communications Board, "Establishment of Ionospheric Stations," March 22, 1944, 1, LAC, RG 24, volume 3414, file 468-4-1.

25. F. T. Davies to Officer-in-Charge, Operational Intelligence Centre, memo, February 21, 1944, LAC, RG 24 volume 4058, file N.S. 1078-13-8.

26. Central Radio Bureau, "Minutes of Discussion on Ionospheric Problems," March 25, 1944, 6, LAC, RG 24, volume 4058, file N.S. 1078-13-8.

27. On tropicality in relation to temperateness, see David Arnold, *The Problem of Nature: Environment, Culture and European Expansion* (Blackwell, 1996); on nordicity, see Patricia Fara, "Northern Possession: Laying Claim to the Aurora Borealis," *History Workshop Journal*, no. 42 (1996): 37–57.

28. Stephen Greenblatt, *Marvelous Possessions: The Wonder of the New World* (University of Chicago Press, 1992), 22–23.

29. Alexander von Humboldt, *Cosmos: A Sketch of a Physical Description of the Universe*, volume 1 (Longman, 1868), 3.

30. Arnold, *The Problem of Nature*, 10.

31. Humboldt, *Cosmos*, 13–14. On Humboldt's influence in British North America, see Suzanne Zeller, "Humboldt and the Habitability of Canada's Great Northwest," *Geographical Review* 96, no. 3 (2006): 382–398.

32. Fara, "Northern Possession," 38; Karen Oslund, "Imagining Iceland: Narratives of Nature and History in the North Atlantic," *British Journal for the History of Science* 35, no. 3 (2002): 313–334.

33. Winston Churchill, *The Second World War*, volume II: Their Finest Hour (Cassell, 1949), 259.

34. On the role of HF-DF, see Kathleen Broome Williams, *Secret Weapon: US High-Frequency Direction Finding in the Battle of the Atlantic* (Naval Institute Press, 1996).

35. See Kurt F. Jensen, *Cautious Beginnings: Canadian Foreign Intelligence, 1939–51* (University of Washington Press, 2009).

36. The auroral zones are the oval-shaped zones of auroral activity that encircle the North and South geomagnetic poles. They contain both the visible aurora and the radio aurora responsible for disruptions of short-wave radio transmissions.

37. "Biographical notes," Frank Davies Papers.

38. Davies and Meek, "Report to CRWPC."

39. For the case of nuclear research, see Robert Bothwell, *Nucleus: The History of Atomic Energy of Canada Limited* (University of Toronto Press, 1988); on the more general anxiety, see Bothwell, *Alliance and Illusion: Canada and the World, 1945–1984* (University of British Columbia Press, 2007).

40. Canadian Radio Wave Propagation Committee (CRWPC), "The Application of Ionospheric Measurements," 1944, 13, LAC, RG 24-C-1, file 9147–2-1.

41. D. A. West has pointed to the more general use of this argument from uniqueness in Canadian Northern scientific research. See D. A. West, "Re-searching the North in Canada: An Introduction to the Canadian Northern Discourse," *Journal of Canadian Studies* 26, no. 2 (1991). Also see Suzanne Zeller, "The Colonial World as Geological Metaphor: Strata(gems) of Empire in Victorian Canada," *Osiris* 15 (2000): 85–107.

42. Frank T. Davies, "DRTE," 5; CRC.

43. See Shelagh D. Grant, "Myths of the North in the Canadian Ethos," *Northern Review*, no. 3/4 (1989): 15–41; Grant, "Northern Nationalists: Visions of a New North, 1940–1950," in *For Purposes of Dominion: Essays in the Honor of Morris Zaslow*, ed. Kenneth Coates and William Morrison (Captus University Publications, 1989). Also see Carl Berger, "The True North Strong and Free," in *Nationalism in Canada*, ed. Peter Russell (McGraw-Hill, 1966); Sherrill Grace, *Canada and the Idea of North* (McGill-Queen's University Press, 2007).

44. See Thomas C. Keefer, *Philosophy of Railroads and Other Essays* (University of Toronto Press, 1972); Daniel Francis, *National Dreams: Myth, Memory, and Canadian History* (Arsenal Pulp Press, 1997); A. A. den Otter, *The Philosophy of Railways: The Transcontinental Railway Idea in British North America* (University of Toronto Press, 1997).

45. These would form part of postwar defense cooperation that included airfields, weather stations, and plans for continental air defense. See William R. Willoughby, *The Joint Organizations of Canada and the United States* (University of Toronto Press, 1979), 129; Richter, *Avoiding Armageddon: Canadian Military Strategy and Nuclear Weapons, 1950–63* (University of British Columbia Press, 2002), 20.

46. Quoted in Clyde Sanger, *Malcolm MacDonald: Bringing an End to Empire* (McGill-Queen's University Press, 1995), 237.

47. Quoted in ibid., 239.

48. See Grant, "Northern Nationalists."

49. Norman Robertson, quoted in J. L. Granatstein, *A Man of Influence: Norman A. Robertson and Canadian Statecraft, 1929–68* (Deneau, 1981), 118.

50. Escott Reid, "The United States and Canada: Domination, Cooperation, Absorption," quoted in Denis Smith, *Diplomacy of Fear: Canada and the Cold War, 1941–1948* (University of Toronto Press, 1988), 18. The anxiety about Canadian dependency touched many areas and shaped the "peripheral dependence" theory starting in the 1950s. For an overview, see David B. Dewitt and John J. Kirton, *Canada as a Principal Power: A Study in Foreign Policy and International Relations* (Wiley, 1983). For a criticism in the area of defense strategy, see Andrew Richter, *Avoiding Armageddon*, Introduction.

51. Carl Berger, "Canadian Nationalism," in *Interpreting Canada's Past*, volume 2: *Post-Confederation*, ed. J. M. Bumsted (Oxford University Press, 1993); Grant, "Myths of the North in the Canadian Ethos"; Renée Hulan, *Northern Experience and the Myths of Canadian Culture* (McGill-Queen's University Press, 2002); Grace, *Canada and the Idea of North*.

52. Quoted in Sanger, *Malcolm MacDonald*, 239.

53. J. T. Wilson, "Defence Research in the Canadian Arctic (with Special Consideration for Geophysical Research)," February 27, 1948, 1, LAC, RG 2, volume 251, file R-100-D 1948–1949, volume 4.

54. Minutes of Canadian Radio Wave Propagation Committee, Ionosphere Sub-Committee, October 5, 1945, 3, LAC, RG 24, volume 3414, file 468-3-3.

55. F. T. Davies, "The Sector Principle in Polar Claims," February 11, 1947, LAC, RG 2/18, volume 46, file A-25. US challenges to Canadian sovereignty in the North had always rested in large part on the principle of "effective occupation," whereby a nation proved its sovereignty over a region by occupying it through settlement or through the coercive organs of the state. Unable to prove effective occupation in the northernmost areas, Canada had claimed sovereignty over its sparsely populated Arctic islands by invoking the "Sector Principle," which granted Canada the entire area bounded by lines of longitude running from its easternmost and westernmost points to the North Pole.

56. "Notes on Discussion of Ionospheric Measurement and Radio Transmission Work of Laboratories at London, Sydney, and Washington, 30 December 1942 to 9 January 1943," January 12, 1943, 3, LAC, RG 24 volume 4058, file NSS 1078–13–8, volume 1.

57. Canadian Radio Wave Propagation Committee, "Minutes of First Meeting," April 10, 1944, 1, LAC, RG 24, volume 3413, file 468-3-2, volume 1; Frank T. Davies and James C. Scott, Canadian Radio Wave Propagation Committee, "Information Bulletin No. 5," July 18, 1946, 1, LAC, RG 24, volume 3413, file HQ 468-3-1.

58. J. H. Meek, "A Proposed Frequency Prediction Service for Canada," January 18, 1946, 1, LAC, RG 24, volume 3414, file 468–3-3.

59. Minutes of Canadian Radio Wave Propagation Committee, February 28, 1947, 7; LAC, RG 24, volume 3413, file HQ 468–3-1 (emphasis added).

60. This way of approaching radio problems—through physical and dynamical investigations rather than statistical correlations—would gain momentum after the war. The Norwegian Defense Research Establishment (NRDE) adopted the Radio Physics Laboratory's more "scientific" approach to the problems of Northern radio communications after a visit to the RPL in 1955. See Olav Wicken, "Space Science and Technology in the Cold War: The Ionosphere, the Military and Politics in Norway," *History and Technology* 13, no. 3 (1997): 207–229.

61. This paralleled work in seismology. See Fa-ti Fan, ""Collective Monitoring, Collective Defense": Science, Earthquakes, and Politics in Communist China," *Science in Context* 25, no. 1 (2012): 127–154.

62. Minutes of Canadian Radio Wave Propagation Committee, February 28, 1947, 7.

63. Liza Piper, "Introduction: The History of Circumpolar Science and Technology," *Scientia Canadensis* 33, no. 2 (2010): 1.

Chapter 2

1. Anxiety about control is a common theme in the literature on scientific instrumentation. See, for example, Sharon Traweek, *Beamtimes and Lifetimes: The World of High Energy Physicists* (Harvard University Press, 1988); Peter Galison, *Image and Logic: A Material Culture of Microphysics* (University of Chicago Press, 1997).

2. Minutes of Canadian Radio Wave Propagation Committee (CRWPC), February 28, 1947, 6, LAC, RG 24, volume 3413, file HQ 468–3-1.

3. See Harry M. Collins, *Changing Order: Replication and Induction in Scientific Practice* (University of Chicago Press, 1992), chapter 3; Joseph O'Connell, "Metrology: The Creation of Universality by the Circulation of Particulars," *Social Studies of Science* 23, no. 1 (1993): 129–173; Bruno Latour, "Give Me a Laboratory and I Will Raise the World," in *Science Observed: Perspectives on the Social Study of Science*, ed. Karin Knorr Cetina and Michael Mulkay (SAGE, 1983); Simon Schaffer, "Late Victorian Metrology and Its Instrumentation: A Manufactory of Ohms," in *The Science Studies Reader*, ed. Mario Biagioli (Routledge, 1999); Peter Galison, *Einstein's Clocks, Poincaré's Maps: Empires of Time* (Norton, 2004).

4. See, for example, Paul N. Edwards, "Meteorology as Infrastructural Globalism," *Osiris* 21 (2006): 229–250; William Rankin, "The Geography of Radionavigation and the Politics of Intangible Artifacts," *Technology and Culture* 55, no. 3 (2014): 622–674.

5. John Tresch explores this dimension of standardization in *The Romantic Machine: Utopian Science and Technology After Napoleon* (University of Chicago Press, 2012), 80.

6. CRWPC, "Standardization of Ionospheric Equipment," May 31, 1945, 4, LAC, RG 24, volume 3413, file HQ 468–3-1.

7. Minutes of the CRWPC, Ionospheric Sub-Committee, October 5, 1945, 2–3, LAC, RG 24, volume 3414, file 468–3-3.

8. J. H. Meek, "A Proposed Frequency Prediction Service for Canada: Report" (January 18, 1946), 4, RG 24, volume 3414, file 468–3-3.

9. Frank T. Davies, "Actual Lecture Memos," 1969, 4, Frank Davies Papers.

10. CRWPC, "Ionospheric Propagation Sub-Committee, Minutes of the First Meeting," May 22, 1944, 1, LAC, RG 24, volume 3414, file 468–3-3.

11. As a generalized record, the A-scan found wider use after World War II in the field of medical imaging. See E. Yoxen, "Seeing with Sound: A Study of the Development of Medical Images," in *The Social Construction of Technological Systems: New Directions in the Sociology and History of Technology*, ed. Wiebe E. Bijker, Thomas P. Hughes, and Trevor Pinch (MIT Press, 1987).

12. The use of this system at Canadian ionospheric stations at the end of World War II is documented in the transcript of the CRWPC Conference, October 22–27, 1945, 27, LAC, RG 24, volume 3412.

13. CRWPC, Ionospheric Propagation Sub-Committee, "Minutes of the First Meeting," May 22, 1944, 2, LAC, RG 24, volume 3414, file 468–3-3.

14. See Yeang, "From Mechanical Objectivity to Instrumentalizing Theory: Inventing Radio Ionospheric Sounders," 195.

15. T. R. Gilliland, who introduced the automatic ionosonde in this form, claimed to do so in order to save time and labor, as well as to capture the details of an often rapidly changing ionosphere. See "Note on a Multifrequency Automatic Recorder of Ionosphere Heights," *Bureau of Standards Journal of Research* 11 (1933): 561–566.

16. Dominion Bureau of Statistics, *The Canada Year Book 1955: The Official Statistical Annual of the Resources, History, Institutions, and Social and Economic Conditions of Canada* (Ottawa: Queen's Printer, 1955), 145.

17. L. H. Wylie, "Memorandum: Proposed Establishment of an R.W.P. Army Station," October 11, 1945, LAC.

18. "Appendix 'B', Minutes of the 13th Meeting of Ionospheric Sub-Committee, Work of Scientific Staff C.R.W.P.C.," January 18, 1946, 2, LAC, RG 24, volume 3414, file 468–3-3.

19. J. H. Meek, "A Historical Outline of Forecasting Methods," paper presented at the AGARD EPC Symposium on Ionospheric Forecasting, St. Jovite, Quebec, September 2, 1969.

20. Robert Marc Friedman, *Appropriating the Weather: Vilhelm Bjerknes and the Construction of a Modern Meteorology* (Cornell University Press, 1989), 3.

21. CRWPC, *Instructions for Observers: Canadian Ionospheric Stations* (Canadian Radio Wave Propagation Committee, July 1944), 2, 4 (emphasis added).

22. NHCO/OA, Tenth Fleet, ASM, BOX 6. HF/DF memo, August 11, 1942, quoted in Kathleen Broome Williams, *Secret Weapon: US High-Frequency Direction Finding in the Battle of the Atlantic* (Naval Institute Press, 1996), 186.

23. Frank T. Davies, Memorandum to Vice-Director General, Defence Research Board, "Assignment of Personnel of Armed Services, and Department of Transport to the Radio Propagation Laboratory," July 16, 1947, 2, LAC, RG 24, volume 83–84/167, file 7401–1710; Frank T. Davies, Memorandum to Director of Signals Division, Naval Service Headquarters, November 4, 1946, LAC, RG 24, volume 83–84/167, file 7401–1710.

24. Davies, Memorandum to Director of Signals Division, November 4, 1946 (emphasis added).

25. P. Whitney Lackenbauer and Matthew Farish, "The Cold War on Canadian Soil: Militarizing a Northern Environment," *Environmental History* 12, no. 4 (2007): 920–950.

26. G. K. Davies, *An Annotated Bibliography of Unclassified Reports Issued by Defence Research Northern Laboratories, 1947–1965* (Ottawa: Defence Scientific Information Service, 1969), 4.

27. J. T. Wilson, "Defence Research in the Canadian Arctic (with Special Consideration for Geophysical Research)," February 27, 1948, 14, LAC, RG 2, volume 251, file R-100-D 1948–1949, volume 4.

28. D. A. Worth, "Memorandum to Administrative Deputy, Defence Research Board," July 23, 1947, LAC, RG 24, volume 83–84/167, file 7401–1710.

29. F. T. Davies, "Assignment of Personnel," 1.

30. Ibid.

31. Draft, Secret Organization Order 234, "Formation of 1 Ionospheric Recording Unit (IRU), Torbay, Nfld.," October 18, 1944, LAC, RG 24, volume 3414, file 468–4-2, 1.

32. A. M. Haig, Memorandum, "Establishment: No.1 Ionospheric Recording Unit, Torbay, Nfld.," March 12, 1945, LAC, RG 24, volume 3414, file 468–4-2.

33. Confidential Memorandum from G. M. Fawcett to D of Ph. & M, "Photographic Equipment for No. 1 Ionosphere Recording Unit," April 4, 1945, LAC, RG 24, volume 3414, file 468–4-2.

34. Draft, Secret Organization Order 234, May 10, 1945, 1 (emphasis added).

35. Ibid., 1.

36. Ibid., 2.

37. Combined Communications Board, Wave Propagation Committee, "Standardization of Characteristics of Ionospheric Recorders," February 19, 1946, 1, LAC, RG 24, volume 3413, file HQ 468–3-2, v. 1.

38. Frank T. Davies and James C. Scott, "Information Bulletin No. 5," July 18, 1946, 2, LAC, RG 24, volume 3413, file HQ 468–3-1.

39. Letter from E. E. Stevens to H. Hutchison, June 13, 1949, 1, LAC, RG 12, volume 1641, file 6802–133.

40. Confidential Memorandum from S. R. Burbank (CAS) to CRWPC (Navy), "Ionospheric Recording Equipment, No. 1 IRU, St. John's, Newfoundland," May 10, 1945, LAC, RG 24, volume 3415, 3416.

41. CRWPC, "Report on CWPC Meeting, May 16, 1945, at Washington D.C.," May 25, 1945, 3, LAC, RG 24, volume 3414, file 468-3-1.

42. Ibid.

43. Memorandum from W. G. L. Flagler, RCNVR, to R2, "Subject—Spares Establishment—Ionospheric Equipment," August 3, 1945, LAC, RG 24, volume 83–84/167, Box 3168, file 7401–1730.

44. Ibid.

45. CRWPC, "Information Bulletin No. 5—Radio Wave Propagation Measurement Program," July 18, 1946, LAC, RG 24, volume 3413, file HQ 468–3-1, 2.

46. J. H. Meek, "Sporadic Ionization at High Latitudes," *Journal of Geophysical Research* 54, no. 4 (1949): 339–345 (emphasis added).

47. Gilliland, "Note on a Multifrequency Automatic Recorder of Ionosphere Heights," 561. Also see Theodore R. Gilliland, "Multifrequency Ionosphere Recording and Its Significance," *Proceedings of the Institute of Radio Engineers* 23 (1935): 1076–1101.

48. CRWPC, *Instructions for Observers*, 2.

49. Minutes of CRWPC, February 28, 1947, 2, LAC, RG 24, volume 3413, file HQ 468–3-1.

50. Combined Communications Board, Wave Propagation Committee, "Standardization of Characteristics of Ionospheric Recorders," February 19, 1946.

51. See, for example, Collins, *Changing Order*, chapter 3; Schaffer, "Late Victorian Metrology"; O'Connell, "Metrology"; Latour, "Give Me a Laboratory and I Will Raise the World."

52. Peter Galison has challenged this role of circulating instruments and people in vouchsafing replication. Peter Galison, "Material Culture, Theoretical Culture and Delocalization," in *Companion to Science in the Twentieth Century*, ed. John Krige and Dominique Pestre (Routledge, 2003). For the example of technology, see Ken Alder, "Making Things the Same: Representation, Tolerance and the End of the *Ancien Regime* in France," *Social Studies of Science* 28, no. 4 (1998): 499–545.

53. I borrow the term "conditions of instrumentality" from Galison ("Material Culture, Theoretical Culture and Delocalization," 671).

54. On the history of worldwide meteorological standardization, see Edwards, "Meteorology as Infrastructural Globalism."

55. "Notes on Discussion of Ionospheric Measurements and Radio Transmission Work of Laboratories at London, Sydney, and Washington, December 30, 1942 to January 9, 1943," January 12, 1943, LAC, RG 24, volume 4058, file NSS 1078–13-8, volume 1, 6.

56. Minutes of CRWPC, February 28, 1947, 2.

57. Transcript of CRWPC Conference, October 22–27, 1945, 26, LAC, RG 24, volume 3412.

58. One American researcher at the conference called the complexity of Canadian records "the most difficult problem in the world, from an ionosphericist's point of view." See "Interpretation of Records," Discussion led by Mr. H. W. Wells, Canadian Radio Wave Propagation Conference, Ottawa, October 22–27, 1945, 59, LAC, RG 24, volume 3412.

59. This position mirrored an ideal put forward during World War II, where, "if two forecasters were given copies of one manual describing a forecast method and placed in separate rooms with current data, they would make identical forecasts." See Irving I. Gringorten, "Methods of Objective Weather Forecasting," *Advances in Geophysics* 2 (1955): 57–92.

60. Transcript of CRWPC Conference, October 22–27, 1945, 26, LAC, RG 24, volume 3412 (emphasis added).

61. Ibid., 27.

62. Minutes of CRWPC, February 28, 1947, 3.

63. Ibid., 6.

64. Frank T. Davies and J. C. Scott, "Radio Wave Propagation Measurement Program." Canadian Radio Wave Propagation Committee, Information Bulletin No. 5, July 18, 1946, Appendix B, LAC, RG 24, volume 3413, HQ 468–3-1.

65. Ibid. Michael Gordin has examined the role of a traveling calibration party in a quite different context—that of Mendeleev's educational reforms in Imperial Russia. See Michael D. Gordin, "Making Newtons: Mendeleev, Metrology, and the Chemical Ether," *Ambix* 45, no. 2 (1998): 96–115.

66. Minutes of CRWPC, February 28, 1947, 6.

67. Letter from E. E. Stevens to W. B. Smith, February 15, 1949, LAC, RG 12, volume 1641, file 6802–133.

68. Telegram from E. E. Stevens to the Controller of Radio, February 25, 1949, LAC, RG 12, volume 1641, file 6802–133.

69. Memo from E. E. Stevens to Controller of Radio, March 15, 1949, LAC, RG 12, volume 1641, file 6802–133.

70. Minutes of the CRWPC, March 15, 1949.

Chapter 3

1. In 1999, a third territory, Nunavut, was created from part of the Northwest Territories.

2. F. H. Collins, *Radio in the Yukon Territory: A brief presented to the Royal Commission on Broadcasting*, April 1956, 4–5, LAC, RG 41, volume 127, file 5.

3. *Current Affairs for the Canadian Forces* 7, no. 4 (1954).

4. The literature on scientific representations is extensive. Some representative works include: Bruno Latour, *Science in Action: How to Follow Scientists and Engineers Through Society* (Harvard University Press, 1987); Caroline Jones and Peter Galison, eds., *Picturing Science, Producing Art* (Routledge, 1998); Michael E. Lynch and Steve Woolgar, eds., *Representation in Scientific Practice* (MIT Press, 1990); Peter Taylor and Ann Blum, "Pictorial Representation in Biology," *Biology and Philosophy* 6, no. 2 (1991): 125–134; Michael Lynch, "Discipline and the Material Form of Images: An Analysis of Scientific Visibility," *Social Studies of Science* 15, no. 1 (1985): 37–66; Martin J. Rudwick, "The Emergence of a Visual Language for Geological Science 1760–1840," *History of Science* 14 (1976): 149–195. The field of laboratory studies is similarly prolific. For a sampling, see Bruno Latour and Steve Woolgar, *Laboratory Life: The Social Construction of Scientific Facts* (Princeton University Press, 1979); Michael Lynch, *Art and Artifact in Laboratory Science: A Study of Shop Work and Shop Talk in a Research Laboratory* (Routledge, 1985); Steven Shapin and Simon Schaffer, *Leviathan and the Air-Pump: Hobbes, Boyle, and the Experimental Life* (Princeton University Press, 1985); Steven Shapin, "The House of Experiment in Seventeenth-Century

England," *Isis* 79, no. 3 (1988): 373–404; Sharon Traweek, *Beamtimes and Lifetimes: The World of High Energy Physicists* (Harvard University Press, 1988).

5. Ian Hacking, "The Self-Vindication of the Laboratory Sciences," in *Science as Practice and Culture*, ed. Andrew Pickering (University of Chicago Press, 1992).

6. On the "image tradition" in modern microphysics, see Peter Galison, *Image and Logic: A Material Culture of Microphysics* (University of Chicago Press, 1997).

7. Latour and Woolgar, *Laboratory Life*.

8. On the sciences of the eye, see Lorraine Daston, "Objectivity and the Escape from Perspective," *Social Studies of Science* 22, no. 4 (1992): 597–618; Lorraine Daston and Peter Galison, *Objectivity* (Zone, 2007).

9. On ionospheric research straddling "experiment" and "observation," see Chen-Pang Yeang, "From Mechanical Objectivity to Instrumentalizing Theory: Inventing Radio Ionospheric Sounders," *Historical Studies in the Natural Sciences* 42, no. 3 (2012): 190–234; Yeang, *Probing the Sky with Radio Waves: From Wireless Technology to the Development of Atmospheric Science* (University of Chicago Press, 2015). Simon Schaffer has also argued that the astrophysical observatories of Victorian Britain combined practices of watching and recording with a lab-like material culture that blurred the line between observation and experiment. See Simon Schaffer, "Where Experiments End: Tabletop Trials in Victorian Astronomy," in *Scientific Practice: Theories and Stories of Doing Physics*, ed. Jed Z. Buchwald (University of Chicago Press, 1995).

10. Daston and Galison, *Objectivity*, chapter 1.

11. The global aspirations of ionospheric research, and the networks designed to realize them, are discussed in chapter 2.

12. When high-frequency radio waves enter the ionosphere over most sections of the world, they are resolved into two separate rays of different polarizations. One of these waves—the ordinary wave—behaves as if there were no magnetic field present. The other—the extraordinary wave—has refractive properties that are a function not merely of the electron density at the point of refraction but also of the strength of the magnetic field at that point. Under very special circumstances associated with high geomagnetic latitudes, a third component called the magneto-ionic component might also appear.

13. For a discussion of the *Typus*, see Daston and Galison, *Objectivity*, 167.

14. Londa Schiebinger, "Skeletons in the Closet: The First Illustrations of the Female Skeleton in Eighteenth-Century Anatomy," *Representations* 14 (1986): 42–82.

15. W. J. G. Beynon and G. M. Brown, eds., *IGY Instruction Manual*, volume 1 (Pergamon, 1956), 111.

16. The authors of the *IGY Instruction Manual* would admit that the standard scaling techniques had proved "far from satisfactory for world-wide use"—see ibid., page 111. The choice of these records was due partly to the geographical distribution of the laboratories (primarily American and European) that had developed the interpretation practices, and partly because of the simplicity of the records. See J. A. Ratcliffe, "The Formation of the Ionosphere: Ideas of the Early Years (1925–1955)," *Journal of Atmospheric and Terrestrial Physics* 36 (1974): 2167–2181.

17. E. E. Stevens, *The Complexity of High Latitude Ionograms* (Ottawa: Defence Research Telecommunications Establishment, May 1961), 1.

18. Brown and Beynon, 105 (emphasis added).

19. "Interpretation of Records." Discussion led by Mr. H. W. Wells, Canadian Radio Wave Propagation Conference, Ottawa, October 22–27, 1945, 59, LAC, RG 24, volume 3412. Although the document is not dated, the context establishes the discussion taking place in the late summer of 1945.

20. On September 30, 1955, for example, an entire day's soundings (96 ionograms) at Barrow, Alaska had produced only one reliable critical frequency for the F2 layer, and none at all for the E and F1 layers. R. W. Knecht, "Statistical Results and Their Shortcomings Concerning the Ionosphere Within the Auroral Zone," *Journal of Atmospheric and Terrestrial Physics* Special Supplement (1957): 109–119.

21. J. H. Meek, "Sporadic Ionization at High Latitudes," *Journal of Geophysical Research* 54, no. 4 (1949): 339–345.

22. J. H. Meek and C. A. McKerrow, *Ionosphere Observer's Instruction Manual* (Ottawa: Defence Research Telecommunications Establishment, February 15, 1951), Section 4.

23. Knecht, "Statistical Results and Their Shortcomings Concerning the Ionosphere Within the Auroral Zone," 115 (emphasis added).

24. J. H. Meek, "Polar Disturbances," in *Polar Atmosphere Symposium, Proceedings of the Symposium Held 2–8 July, 1956 in Oslo*, ed. Sir Edward Appleton, Special Supplement (Pergamon, 1957), 121–122.

25. J. H. Meek and A. G. McNamara, "Magnetic Disturbances, Sporadic E, and Radio Echoes Associated with the Aurora," *Canadian Journal of Physics* 32 (1954): 326–329.

26. J. H. Meek, "A Historical Outline of Forecasting Methods," paper presented at AGARD EPC Symposium on "Ionospheric Forecasting," St. Jovite, Canada, September 2, 1969, 6.

27. Meek, "Polar Disturbances," 125.

28. Knecht, "Statistical Results and Their Shortcomings Concerning the Ionosphere Within the Auroral Zone," 117.

29. W. J. G. Beynon and G. M. Brown, eds., *IGY Instruction Manual: The Ionosphere, Part I, Ionospheric Vertical Soundings* (Pergamon, 1957), 105.

30. Ibid.

31. On the distinction between serial and parallel graphic products, see Michael Lynch, "Discipline and the Material Form of Images."

32. J. C. W. Scott, "Memorandum on Canadian Space Research Program," November 19, 1958, 1, Frank Davies Papers.

33. D. J. Goodspeed, *A History of the Defence Research Board of Canada* (Ottawa: Queen's Printer, 1958), 195.

34. Harold Innis, *The Fur Trade in Canada* (University of Toronto Press, 1970), 393.

35. Harold Innis, *Changing Concepts of Time* (University of Toronto Press, 1952), 18–19.

36. CBC, "CBC Northern Service," June 22, 1965, 1, LAC, RG 41, volume 128, file 5–4.

37. Andrew Cowan, "Consideration of Broadcast Coverage for the Yukon and Northwest Territories," December 12, 1955, 4, LAC, RG 41, volume 127, file 5–2.

38. T. Ringerside, "Communications Facilities and User Problems in the Canadian Arctic," *Polar Record* 14, no. 90 (1968): 305–314.

39. CBC, "CBC Northern Service," June 22, 1965, 1.

40. Minutes of 4th Annual Meeting of Station Managers, January 13–17, 1964, 16, LAC, RG 41, volume 127, file 5–3.

41. Canadian Broadcasting Corporation, "Northern Service," April 1963, 1, LAC, RG 85, volume 145, file 1003–17, part 9 (emphasis added).

42. Louis-Edmond Hamelin, *Nordicité Canadienne* (Hurtubise, 1975), 35. Although Hamelin's final system was not presented until the 1970s, there is evidence that the underlying principles of nordicity along with its various criteria were worked out in the late 1950s and the early 1960s.

43. Hamelin's first sketch of this system was presented in 1964 in an attempt to arrive at an accurate delimitation of the Canadian North. Communications formed one of ten criteria in this emerging system. See Louis-Edmond Hamelin, "Essai De Régionalisation Du Nord Canadien," *North* (1964): 16–19.

44. Louis-Edmond Hamelin, *About Canada: The Canadian North and Its Conceptual Referents* (Ottawa: Department of the Secretary of State, 1988), 21.

45. On regimes of perceptibility, see Michelle Murphy, *Sick Building Syndrome and the Problem of Uncertainty: Environmental Politics, Technoscience, and Women Workers* (Duke University Press, 2006), 12.

Chapter 4

1. Robert Scott, "What Happened on the Night of July 8?" *Honolulu Star-Bulletin*, July 28, 1962. For a detailed analysis of the effects of the blast, see Charles N. Vittitoe, Did High-Altitude EMP Cause the Hawaiian Streetlight Incident? System Design and Assessment Note 31, Sandia National Laboratories, June 1989.

2. P. Whitney Lackenbauer and Matthew Farish, "The Cold War on Canadian Soil: Militarizing a Northern Environment," *Environmental History* 12, no. 4 (2007): 920–950.

3. The anxiety over them helped create world-spanning sonar and meteorological networks. See Michael Aaron Dennis, "Postscript: Earthly Matters: On the Cold War and the Earth Sciences," *Social Studies of Science* 33, no. 5 (2003): 809–819; Paul N. Edwards, "Meteorology as Infrastructural Globalism," *Osiris* 21 (2006): 229–250.

4. Walter A. McDougall, "Space-Age Europe: Gaullism, Euro-Gaullism, and the American Dilemma," *Technology and Culture* 26, no. 2 (1985): 179–203; Howard E. McCurdy, *Space and the American Imagination* (Johns Hopkins University Press, 2011); Roger D. Launius, "Perfect World, Perfect Societies: The Persistent Goal of Utopia in Human Spaceflight," *Journal of the British Interplanetary Society* 56 (2003): 338–349; Asif Siddiqi, "National Aspirations on a Global Stage: Fifty Years of Spaceflight," in *Remembering the Space Age*, ed. Steven Dick (NASA, 2008); James Hansen, "The Great Leap Upward: China's Human Spaceflight Program and Chinese National Identity," in *Remembering the Space Age*, ed. Steven Dick (NASA, 2008).

5. Lackenbauer and Farish, "The Cold War on Canadian Soil."

6. Martin W. Lewis and Kären Wigen, *The Myth of Continents: A Critique of Metageography* (University of California Press, 1997), 163.

7. See Matthew Farish, *The Contours of America's Cold War* (University of Minnesota Press, 2010), 64.

8. Lackenbauer and Farish, "The Cold War on Canadian Soil: Militarizing a Northern Environment," 921. On the normalization of the abnormal geography of World War II in the United States, see Gretchen Heefner, *The Missile Next Door: The Minuteman in the American Heartland* (Harvard University Press, 2012), 66. On the links between wartime regions and Cold War "area studies," see Lewis and Wigen, *The Myth of Continents*, 163.

9. Lloyd V. Berkner, "Electronic Design for Low Temperature." Informal talk given at the technical program of the Subcommittee on the Prevention of Deterioration of

Electrical and Electronic Materials, February 10, 1948, National Academy of Sciences, Washington D.C., 3.

10. Ibid. On strategic gaming at RAND, see Sharon Ghamari-Tabrizi, *The Worlds of Herman Kahn: The Intuitive Science of Thermonuclear War* (Harvard University Press, 2005).

11. Dennis, "Postscript: Earthly Matters: On the Cold War and the Earth Sciences," 809; Lackenbauer and Farish, "The Cold War on Canadian Soil," 943.

12. See Ronald Doel, "Earth Sciences and Geophysics," in *Science in the Twentieth Century*, ed. John Krige and Dominique Pestre (Harwood, 1997).

13. Jacob Darwin Hamblin, "The Navy's "Sophisticated" Pursuit of Science," *Isis* 93, no. 1 (2002): 1–27.

14. Laura A. Bruno, "The Bequest of the Nuclear Battlefield: Science, Nature, and the Atom during the First Decade of the Cold War," *Historical Studies in the Physical and Biological Sciences* 33, no. 2 (2003): 237–260.

15. Technological Capabilities Panel, *Meeting the Threat of Surprise Attack* (Science Advisory Committee, February 14, 1955).

16. See Gregory Good, "Sydney Chapman: Dynamo Behind the International Geophysical Year," in *Globalizing Polar Science: Reconsidering the International Polar and Geophysical Years*, ed. Roger D. Launius, James Rodger Fleming, and David H. DeVorkin (Palgrave Macmillan, 2010); Allan A. Needell, *Science, Cold War and the American State* (Amsterdam: Harwood Academic Publishers, 2000); Spencer R. Weart, "Global Warming, Cold War, and the Evolution of Research Plans," *Historical Studies in the Physical and Biological Sciences* 27, no. 2 (1997): 319–356; Ronald Doel, "Constituting the Postwar Earth Sciences: The Military's Influence on the Environmental Sciences in the USA After 1945," *Social Studies of Science* 33, no. 5 (2003): 635–666; Oreskes and Doel, "The Physics and Chemistry of the Earth," 555; John Krige, "Atoms for Peace, Scientific Internationalism, and Scientific Intelligence," *Osiris* 21, no. 1 (2006): 161–181. For a more general overview of the IGY in the context of the International Polar Years, see Launius, Fleming, and DeVorkin, *Globalizing Polar Science*.

17. Robert W. Buchheim, *Space Handbook: Astronautics and Its Applications*. (Random House, 1958), 9.

18. For a general discussion of this "reliability crisis" and specific responses, see Edward Jones-Imhotep, "Disciplining Technology: Electronic Reliability, Cold-War Military Culture and the Topside Ionogram," *History and Technology* 17, no. 2 (2000): 125–175. On the impact on the visual culture of Cold War electronics, see Jones-Imhotep, "Icons and Electronics," *Historical Studies in the Natural Sciences* 38, no. 3: 405–450.

19. Headquarters Air Force Systems Command, History Office, *Toward New Horizons: Science, the Key to Air Supremacy* (Andrews Air Force Base, 1992), xi, 6.

20. Keith Henney, ed., *Reliability Factors for Ground Electronic Equipment* (McGraw-Hill, 1956), 1-1.

21. Radio Physics Laboratory, "Annual Report of Progress, 1958," 1, LAC, RG 24, volume 24043, file 95H-36.

22. Matthias Heymann et al., "Exploring Greenland: Science and Technology in Cold War Settings," *Scientia Canadensis* 33, no. 2: 11–42.

23. Radio Physics Laboratory, "Annual Report of Progress, 1958," 2.

24. See ibid. In 1958, 40 percent of PARL radar time would be dedicated to this question.

25. John H. Chapman, "Tentative Proposals for Research Program for PARL," Memorandum from John Chapman to Chief Superintendent, DRTE, August 15, 1958, 2, LAC, RG 41, accession J43, volume 4, file 4–92; John H. Chapman, "Estimates 1960–61," Memorandum from John Chapman to Administrative Officer, DRTE, April 24, 1959, 1, LAC, RG 31, accession J43, volume 4, file 4–9.

26. John H. Chapman, R. K. Brown, and W. Heikkila, "The DRTE Satellite and Rocket Program." Paper presented at the 11th DRB Symposium, December 9, 1959, 1, LAC, RG 31, accession J43, file 1–33.

27. For an overview of the Soviet program, see Asif A. Siddiqi, *Sputnik and the Soviet Space Challenge* (Gainesville: University Press of Florida, 2003), *Challenge to Apollo: The Soviet Union and the Space Race, 1945–1974* (NASA History Office, 2010); Siddiqi, *The Red Rockets' Glare: Spaceflight and the Soviet Imagination, 1857–1957* (Cambridge University Press, 2010). On the wider cultural resonance in revolutionary Russia, see Asif A. Siddiqi, "Imagining the Cosmos: Utopians, Mystics, and the Popular Culture of Spaceflight in Revolutionary Russia," *Osiris* 23, no. 1 (2008): 260–288.

28. Directorate of Scientific Information, *The USSR Earth Satellite Project for the International Geophysical Year*, DSI Report no. 8/56 (Ottawa: Directorate of Scientific Research, 1956), 5.

29. John Chapman, Memorandum from John Chapman to Administrative Officer, DRTE.

30. Explorer I, the first successful American satellite, was designed specifically to study the effects of the space environment on spacecraft operating there. It contained three experiments: the first measured temperature extremes on the satellite and in the vicinity of the high-power transmitter; the second measured the impact of micrometeorites; the third was the cosmic-ray experiment of Van Allen. See

Clayton R. Koppes, *JPL and the American Space Program: A History of the Jet Propulsion Laboratory* (Yale University Press, 1982), 86.

31. Directorate of Scientific Information, *The USSR Earth Satellite Project*, 7.

32. Siddiqi, *Challenge to Apollo*, 175.

33. W. J. G. Beynon, "U.R.S.I. And the Early History of the Ionosphere," *Philosophical Transactions of the Royal Society of London Series A* 280, no. 1293 (1975): 47–55.

34. J. C. W. Scott, "Memorandum on Canadian Space Research Program," November 19, 1958, 1, Frank Davies Papers.

35. Letter from John H. Chapman to J. J. Green, August 26, 1958, 1, LAC, RG 41, accession J43, volume 4, file 4–9.

36. C. A. Franklin, "Alouette/ISIS: How It All Began," Address at IEEE International Milestone in Engineering Ceremony, Shirley's Bay, Ottawa, May 13, 1993.

37. J. C. W. Scott, "Meeting on the DRTE Proposal for a Polar Satellite Ionosonde, Room 3D170 Pentagon, 13 November, 1958," Frank Davies Papers.

38. Scott, "Memorandum on Canadian Space Research Program," 1.

39. On the DRB's development of defense strategy, see Andrew Richter, *Avoiding Armageddon: Canadian Military Strategy and Nuclear Weapons, 1950–63* (University of British Columbia Press, 2002).

40. Scott, "Memorandum on Canadian Space Research Program," 1.

41. For comprehensive and recent surveys of space historiography, see Roger D. Launius, "The Historical Dimension of Space Exploration: Reflections and Possibilities," *Space Policy* 16, no. 1 (2000): 23–38; Asif A. Siddiqi, "American Space History: Legacies, Questions, and Opportunities for Future Research," *Critical Issues in the History of Spaceflight*, no. 15 (2006): 433–480; Siddiqi, "Competing Technologies, National(ist) Narratives, and Universal Claims: Toward a Global History of Space Exploration," *Technology and Culture* 51, no. 2 (2010): 425–443.

42. Scott, "Memorandum on Canadian Space Research Program," 1.

43. John E. Jackson, *Alouette-ISIS Program Summary* (National Space Science Data Center, 1986), 14.

44. The language of laboratory for space is common in this period. It emphasized the view of space as an environment. Soviet engineers often refer them as "cosmic laboratories." Already in 1955, the scientific secretary of the USSR Interagency Commission on Interplanetary Travel explained that one of the first tasks would be the creation of an "automatic laboratory for scientific research in cosmic space." See Directorate of Scientific Information, *The USSR Earth Satellite Project for the International Geophysical Year*, 7.

45. J. O. Thomas, "Canadian Satellite: The Topside Sounder Alouette," *Science* 139, no. 3551 (1963): 229–232; John E. Jackson, *Results from Alouette 1, Explorer 20, Alouette 2, and Explorer 31* (National Space Science Data Center, 1988), 37.

46. Jackson, *Alouette-ISIS Program Summary*, 5–6.

47. Siddiqi, *Challenge to Apollo.*

48. A. R. Molozzi, ""Instrumentation of the Ionospheric Sounder Contained in the Satellite 1962 Beta Alpha (Alouette)," in *Space Research IV* (Proceedings of the Fourth International Space Symposium), ed. P. Muller (North-Holland, 1964), 427.

49. Morris Halio, "Improving Electronic Reliability," *IRE Transactions on Military Electronics* 1051, no. 1 (1961): 11–18.

50. At its simplest, the reliability of a system containing n components is expressed as the product of the reliabilities of the individual parts: $R_{\text{system}} = R_1 \times R_2 \times R_3 \times \ldots R_n$.

51. Brigadier General J. M. Colby, "Minutes of the Second AOMC Agencies Conference Regarding Guided Missile Reliability," May 12, 1959, 1, LAC, RG 24, accession 1983–84/049, volume 1664, file 1950–123–7, part 2.

52. Janet Abbate, "Cold War and White Heat: The Origins and Meanings of Packet Switching," in *The Social Shaping of Technology*, ed. Donald MacKenzie and Judy Wajcman (Open University Press, 1999), 357.

53. See Paul Baran, *On Distributed Communications*, volume V: *History, Alternative Approaches, and Comparisons* (RAND Corporation, 1964).

54. Quoted in K. Johnsen, "Senate Vote Emphasizes Proven Weapons," *Aviation Week and Space Technology* 22 (1961): 22.

55. Chester I. Soucy, "The 'Project MATURE' Concept of the Design and Production Requirements for Reliable and Maintainable Military Electronics Equipment." presented to the Third Institute of Radio Engineers Convention, Toronto, October 8, 1958, 5, LAC, accession 1983–84/049, volume 1664, file 1950–123–7, part 2.

56. See Chester I. Soucy, "Reliability Control of Electronic Equipment in Aircraft and Weapons Systems: General and Management Aspects," *Canadian Aeronautics Journal* 3 (1957): 223.

57. Chester I. Soucy, "Some Guiding Principles in Formulating a General Reliability and Maintainability Specification for Electronic Equipment," January 28, 1958, 2–3, LAC, RG 24, file 1950–23–7, part 2.

58. Letter from Chester I. Soucy to Deputy Minister of Defence Production, March 22, 1960, 2, LAC, RG 24, accession 1983–84/049, box 1664, file 1950–123–7, part 3.

59. Soucy, "Project MATURE,'" 5.

60. Ibid.

61. Frank T. Davies, "Actual Lecture Memos," 4, Frank Davies Papers.

62. See Molozzi, "Instrumentation," 427.

63. N. Moody and C. Florida, "Some New Transistor Bistable Elements for Heavy Duty Operation," *IRE Transactions on Circuit Theory* 4, no. 3 (1957): 241–241 (emphasis added).

64. See Julius Lukasiewicz, "Canada's Encounter with High-Speed Aeronautics," *Technology and Culture* 27, no. 2 (1986): 223–261; Donald C. Story and Russell Isinger, "The Origins of the Cancellation of Canada's Avro CF-105 Arrow Fighter Program: A Failure of Strategy," *Journal of Strategic Studies* 30, no. 6 (2007): 1025–1050.

65. Andrew B. Godefroy, *Defence and Discovery: Canada's Military Space Program, 1945–74* (University of British Columbia Press, 2012), 109, note 34.

66. Mar's camera featured a maximum aperture of f/0.71. See J. Mar, "Design Aspects of the Auroral Movie Camera." Defence Research Northern Laboratory Technical Note 35/55, September 6, 1955.

67. DRTE Technical Memorandum no. 308, "First Quarterly Progress Report, Topside Sounder Satellite, for the period ending 31 March 1960," Defence Research Board, Ottawa, March 31, 1960, 2.

68. Molozzi, "Instrumentation," 417.

69. Jackson, *Alouette-ISIS Program Summary*, 18.

70. For a detailed discussion of the techniques and inspection procedures used for the S-27 payload, see G. H. C. Mackie, *Construction Techniques and Inspection Procedures for the S-27 Alouette Satellite Payload*, DRTE Report No. 1119, PCC No. D48–02–04–01 (Defence Research Telecommunications Establishment, Ottawa, 1963).

71. Letter from A. H. Zimmerman to T. K. Glennan, December 1959, United States International Space Programs, 305.

72. According to Canadian military regulations adopted in August 1957, denoted MIL-STD-243, six models of communications and electronic equipment were required: breadboard, experimental, developmental, service test, prototype, and production. See appendix A to Joint Services Electronics Committee memo from W. J. Eastwood to the secretary of PSOC, August 8, 1957, LAC, RG 24, volume 24066, file 9500–12.

73. "Alouette Satellite News Release." July 1962, LAC, RG 24, box 7316, file DRBS 0204–01, part 6.

74. Frank Mee, "SPAR Aerospace, STEM Antenna." Transcript of videotaped presentation, August 22, 1991; National Museum of Science and Technology, Ottawa.

75. Godefroy, *Defence and Discovery*, 105–106.

76. H. R. Warren and J. Mar, "Structural and Thermal Design of the Topside Sounder Satellite," *Canadian Aeronautics and Space Journal* 8, no. 7 (1962): 161–169.

77. John Mar, "Notes on Fifth Meeting of Topside Sounder Working Group at CRPL, 12 October 1960: Appendix C," 1, LAC, RG 24, accession 1983–84/167, box 7315, file DRBS 0204–01 volume 2.

78. Molozzi, "Instrumentation," 423–424.

79. C. A. Franklin, R. J. Bibby, and R. F. Sturrock, "Telemetry and Command Systems for the Canadian Ionospheric Satellite," (Ottawa: Department of National Defence, 1963), 14–15.

80. DRTE Technical Memorandum no. 308, "Third Quarterly Progress Report, Topside Sounder Satellite, for the period ending 31 March 1960," Defence Research Board, Ottawa; September 30, 1960, 15.

81. Ibid., 39.

82. Ibid., 15.

83. DRTE, "Fourth Quarterly Report," 10.

84. DRTE Technical Memorandum no. 348,"Fifth Quarterly Progress Report: Topside Sounder Satellite for the period ending 21 March, 1961," Defence Research Board, Ottawa, March 31, 1961, 10.

85. Ibid., 11.

86. DRTE Technical Memorandum no. 363, "Sixth Quarterly Progress Report: Topside Sounder Satellite for the period ending 30 June, 1961," Defence Research Board, Ottawa, June 30, 1961, 10.

87. DRTE, "Fourth Quarterly Report," 8.

88. DRTE, "Fifth Quarterly Report," 2, 7.

89. DRTE, "Sixth Quarterly Report," 17.

90. DRTE, "Fifth Quarterly Report," 7–9.

91. DRTE, "Sixth Quarterly Report," 9.

92. DRTE Technical Memorandum, "Seventh Quarterly Progress Report: Topside Sounder Satellite for the period ending 30 September, 1961," Defence Research Board, Ottawa, September 30, 1961, 2.

93. Notes on Fifth Meeting of Topside Sounder Working Group at CRPL, October 12, 1960, Appendix C (DRBS 0204–01 volume 2).

94. Author's interview with Colin Franklin, Ottawa, July 17, 1998 (hereafter cited as Franklin interview).

95. John Mar, "Thermal Vacuum Environmental Test Plan." DRTE 0204–01–5 (E), 1.

96. Ibid.

97. Ibid., 1–2.

98. Ibid., 2.

99. It would continue this rate for the first three years of operation. Jackson, *Alouette-ISIS Program Summary*, 19.

100. R. W. Knecht and T. E. Van Zandt, "Some Early Results from the Ionospheric Topside Sounder Satellite." *Nature* 197, no. 4868 (1963): 641–644.

101. G. E. K. Lockwood, "Plasma and Cyclotron Spike Phenomena Observed in Topside Ionograms," *Canadian Journal of Physics* 41, no. 1 (1963): 190–194.

102. John H. Chapman, "Alouette Topside Sounder Satellite—Experiments, Data, and Results," *Journal of Spacecraft and Rockets* 1, no. 6 (1964): 684–686.

103. G. L. Nelms, "Ionospheric Results from the Topside Sounder Satellite Alouette," in *Space Research IV*, 437.

104. See Chapman, "Alouette Topside Sounder Satellite."

105. Jackson, *Alouette-ISIS Program Summary*, 17.

106. Godefroy, *Defence and Discovery*, 114–115.

Chapter 5

1. On the use of computers as a solution to these problems, see Jon Agar, "Making a Meal of the Big Dish: The Construction of the Jodrell Bank Mark 1 Radio Telescope as a Stable Edifice, 1946–57," *British Journal for the History of Science* 27, no. 1 (1994): 3–21; Galison, *Image and Logic*, chapter 5; Robert W. Seidel, "From Factory to Farm: Dissemination of Computing in High-Energy Physics," *Historical Studies in the Natural Sciences* 38, no. 4 (2008): 479–507.

2. See Paul N. Edwards, "Meteorology as Infrastructural Globalism." *Osiris* 21 (2006): 229–250.

3. Jackson, Alouette-ISIS Program Summary, 25. John Krige has examined NASA's increasingly global reach, but his study largely ignores the significance of Alouette in NASA's international programs. See *NASA in the World: Fifty Years of International*

Collaboration in Space, ed. John Krige, Angelina Long Callahan, and Ashok Maharaj (Palgrave, 2013).

4. John E. Jackson, *Results from Alouette 1, Explorer 20, Alouette 2, and Explorer 31* (National Space Science Data Center, 1988), 8; Edwards, "Meteorology as Infrastructural Globalism."

5. Jackson, *Results from Alouette 1*, 1.

6. Walter A. McDougall, *The Heavens and the Earth: A Political History of the Space Age* (Johns Hopkins University Press, 1997), 177.

7. See John Krige, "Embedding the National in the Global: US-French Relationships in Space Science and Rocketry in the 1960s," in *Science and Technology in the Global Cold War*, ed. Naomi Oreskes and John Krige (MIT Press, 2014), 229.

8. Sunny Tsiao, *"Read You Loud and Clear!" The Story of NASA's Spaceflight Tracking and Data Network* (NASA History Division, 2008), xxxiv.

9. See, respectively, Krige et al., *NASA in the World*; Tsiao, *"Read You Loud and Clear!"*

10. On US stipulations about space science and data sharing, see McDougall, *The Heavens and the Earth*, 353. On the place of national interests in the creation of these international space collaborations, see Krige, "Embedding the National in the Global."

11. On the development of this system, see Tsiao, *"Read You Loud and Clear!"*

12. Ibid. About half were dedicated to satellite tracking and the others to human spaceflight.

13. Ibid.

14. Ibid.

15. The single exception to the transfer of tapes to Ottawa was the Singapore station, operated by the British. See C. A. Franklin, R. J. Bibby, and N. S. Hitchcock, "A Data Acquisition and Processing System for Mass Producing Topside Ionograms," *Proceedings of the IEEE* 57, no. 6 (1969): 929–944.

16. Arnold W. Frutkin, *International Cooperation in Space* (Prentice-Hall, 1965), 24–25. Also see Krige, Callahan, and Maharaj, *NASA in the World*, 27–30.

17. For a full discussion, see Walter A. McDougall, *The Heavens and the Earth*, 184–185.

18. Its membership consisted of Argentina, Australia, Belgium, Brazil, Canada, Czechoslovakia, France, India, Iran, Italy, Japan, Mexico, Poland, Sweden, the Union of Soviet Socialist Republics, the United Arab Republic, the United Kingdom of Great Britain and Northern Ireland, and the United States of America.

19. Department of External Affairs, "Outer Space," Memorandum from Department of External Affairs to Permanent Mission of Canada to the United Nations (Permis), New York, January 22, 1959, LAC, RG 25, volume 7840, file 12798–1-40, 1–3.

20. Norman N. A. Robertson, "Memorandum for the Minister: Outer Space," January 23, 1959, LAC, RG 25, volume 7840, file 12798–1-40, 2–3.

21. Godefroy, *Defence and Discovery*, 75, note 13. Early in the space age, it was believed that radiation in the Van Allen belts would make manned spaceflight launches only possible from sites near the poles, where radiation was weaker. See Robert W. Buchheim, *Space Handbook: Astronautics and Its Applications* (Random House, 1958), 24–25.

22. COSPAR was composed of the United States and the Soviet Union (launching countries); Australia, Canada, France, Japan, and the United Kingdom (countries with rocket programs); India, Peru, and South Africa (temporary rotating members engaged in tracking and other aspects of space research); and representatives from nine URSI sections.

23. Robertson, "Memorandum for the Minister: Outer Space."

24. Department of External Affairs, "Outer Space."

25. Memorandum from B. A. Walker to Vice Chairman, Defence Research Board, April 4, 1961, LAC, RG 24, accession 83–84/167, volume 7316, DRBS 0204–01, volume 6.

26. Ernest R. May and Philip Zelikow, *The Kennedy Tapes: Inside the White House during the Cuban Missile Crisis* (Cambridge University Press, 1997), 645.

27. B. A. Walker, Department of National Defence Minute Sheet, April 5, 1961, LAC, RG 24, accession 83–84/167, volume 7316, DRBS 0204–01, volume 6.

28. Letter from J. H. Chapman to J. E. Jackson, March 7, 1961, LAC, RG 24, accession 83–84/167, volume 7316, DRBS 0204–01, volume 6.

29. Ibid.

30. For the details of nuclear isotope sharing, see John Krige, "Atoms for Peace."

31. Letter from J. H. Chapman to John F. Clark (NASA), October 13, 1964, 2, LAC, RG 24, accession 83–84/167, volume 7316, DRBS 0204–01, volume 6.

32. The final arrangement was motivated largely by the activities of Australian scientists who had begun using Alouette data collected from Australian stations without consulting either DRTE or NASA regarding priority. See Letter from Chapman to Clark, October 13, 1964. For details of the NASA arrangement, see Letter from J. H. Chapman to John A. Thomas (University of Melbourne), August 10, 1964, 1, LAC, RG 24, volume 27074, file 9511–40.

33. Letter from J. H. Chapman to Director, Ionospheric Prediction Service, Department of Interior, Australia, November 3, 1965, LAC, RG 24, volume 27074, file 9511–40.

34. On the history of the geological record, see Martin J. Rudwick, "The Emergence of a Visual Language for Geological Science 1760–1840," *History of Science* 14 (1976): 149–195.

35. Lorraine Daston, "The Sciences of the Archive," *Osiris* 27, no. 1 (2012): 156–187.

36. Letter from F. T. Davies to Chairman, Defence Research Board, May 24, 1962, LAC, RG 24, volume 24098, file DRBC 9752–01.

37. Interview with Colin Franklin, July 16, 1998 (hereafter cited as Franklin interview).

38. Jackson, *Results from Alouette 1*, 2.

39. Jackson, *Alouette-ISIS Program Summary*, 15.

40. Franklin, Bibby, and Hitchcock, "A Data Acquisition and Processing System," 935.

41. Franklin interview.

42. Letter from R. H. Dohoo to Director, Telecommunications and Electronics, Department of Transport, October 8, 1964, 2, LAC, RG 24, volume 24098, file DRBC 9752–01.

43. Franklin, Bibby, and Hitchcock, "A Data Acquisition and Processing System," 938.

44. Ibid.

45. Ibid.

46. Franklin, Bibby, and Hitchcock, "A Data Acquisition and Processing System," 944.

47. Frank T. Davies, "Alouette—Her First Year in Orbit," 1963, 2, Frank Davies Papers.

48. The difficulties of generating reliable N(h) profiles are well documented in the ionospheric research literature. For an early attack on the problem, see H. G. Booker and S. L. Seaton, "Relation between actual and virtual ionospheric heights," *Physical Review* 57, no. 2 (1940): 87–94. For later approaches to the problem, see *Radio Science*, volume 2, no. 10 (1967). For a historical analysis of the role of the profiles, see Chen-Pang Yeang, *Probing the Sky with Radio Waves: From Wireless Technology to the Development of Atmospheric Science* (University of Chicago Press, 2015); Aitor Anduaga Egaña, *Wireless and Empire: Geopolitics, Radio Industry, and*

Ionosphere in the British Empire, 1918–1939 (Oxford University Press, 2009), chapter 1 and epilogue.

49. EDSAC was, in fact, the first operational full-scale stored-program computer, and the basis of British priority claims in the invention of the computer. On the history of EDSAC programming, see Martin Campbell-Kelly, "Programming the EDSAC: Early Programming Activity at the University of Cambridge," *Annals of the Journal of Computing* 2 (1980): 7–36. On the problem of calculating electron profiles, with reference to the EDSAC, see K. G. Budden, "The Numerical Solution of Differential Equations Governing Reflexion of Long Radio Waves From the Ionosphere," *Proceedings of the Royal Society of London Series A* 227, no. 1171 (1955): 516–537.

50. Interview with Leroy Nelms, July 15, 1998.

51. J. H. Chapman, "Alouette Satellite," Paper presented at the University of Toronto, November 19, 1962, 10, LAC, RG 31, accession J43, file 1–39.

52. Jackson, *Results from Alouette 1*, 26.

53. Peter Galison has examined a similar separation between data production and data analysis in his study of the bubble chamber. See Galison, *Image and Logic*, 557.

54. This gendered division of labor is pervasive in modern science and technology. See, for example, Jennifer S. Light, "When Computers Were Women," *Technology and Culture* 40, no. 3 (1999): 455–483. On women's involvement in the analysis of star charts, see Pamela Mack, "Straying from Their Orbits: Women in Astronomy in America," in *Women of Science: Righting the Record*, ed. G. Kass-Simon and Patricia Farnes (Indiana University Press, 1993). On "scanner-girls" in nuclear physics, see Galison, *Image and Logic*, chapter 5.

55. DRTE had just put into operation the first transistorized general-purpose computer designed and built in Canada as part of an initiative to foster domestic self-reliance in digital electronics. See John N. Vardalas, *The Computer Revolution in Canada: Building National Technological Competence* (MIT Press, 2001), chapter 3.

56. J. W. Cox, *The Recording and Analysis of Radio Soundings of the Ionosphere from a Satellite* (Ottawa: Defence Research Telecommunications Establishment, 1960)

57. J. H. Chapman, "DRTE Report to the Nineteenth Meeting of the Topside Sounder Working Group," April 19, 1963, 2, LAC, RG 24, accession 83–84/167, volume 7, file 0204–01.

58. L. E. Petrie, "Probability Predictions for Radio Traffic Frequencies, Using a Digital Computer," (Ottawa: Defence Research Telecommunications Establishment, 1960). On the more general state of this research, see Letter from J. H. Morgan to Frank Davies, December 16, 1963, 1, LAC, RG 24, volume 24074, file 9511–40.

59. Cox, "The Recording and Analysis of Radio Soundings of the Ionosphere from a Satellite." Also see E. L. Hagg, E. J. Hewens, and G. L. Nelms, "The Interpretation of Topside Sounder Ionograms," *Proceedings of the IEEE* 57, no. 6 (1969): 949–960; Jackson, *Results from Alouette 1*, 2–3.

60. M. A. Maclean, "DAR Prepares Radar Data for Analysis," *Canadian Electronics Engineering* 4 (1960): 22.

61. L. E. Petrie, *High-Frequency Propagation Predictions Using an IBM 650 Computer* (Ottawa: Defence Research Telecommunications Establishment, 1963).

62. G. E. K. Lockwood, "A Computer-Aided System for Scaling Topside Ionograms," *Proceedings of the IEEE* 57, no. 6 (1969): 986–989.

63. Ibid.

64. Galison, *Image and Logic*, chapter 7.

65. Lockwood, "A Computer-Aided System," 986.

66. George Jull, "History of Informatique at DRTE—CRC 1952–72: Man-Machine Interactions, Data Processing and Storage," 1972, 1, Frank Davies Papers.

67. W. L. Hatton, *Future Communication Systems—Impact of Information Theory* (Ottawa: Defence Research Telecommunications Establishment, 1960).

68. D. C. Coll, *LOGICOM: A Digital Communication System* (Ottawa: Defence Research Telecommunications Establishment, February 1961).

69. R. Hedemark, *Channel Evaluation in a Variable Rate Satellite Communication System* (Ottawa: Defence Research Telecommunications Establishment, May 1965).

70. Researchers cited a number of advantages to the "digital ionogram," among them an increased probability of detection, a constant false alarm rate, a two-level display that didn't require AGC, removal of noise bands from power stations, and the potential for automatic analysis of the digital signal using a digital computer. See D. C. Coll, J. R. Storey, and H. M. Pearce, "Digital Detection of Coded-Pulse Ionosonde Signals," *Proceedings of the IEEE* 53, no. 2 (1965): 188–189.

71. Lockwood, "A Computer-Aided System," 989.

72. Ronald Kline suggests that these concerns, framed around an analogy between humans and machines, dominated cybernetics research above and beyond the creation of cyborgs so central to STS thinking. See "Where Are the Cyborgs in Cybernetics?" *Social Studies of Science* 39, no. 3 (2009): 331–362.

73. Peter Galison, "The Ontology of the Enemy: Norbert Wiener and the Cybernetic Vision," *Critical Inquiry* 21, no. 1 (1994): 228–266.

74. Lockwood, "A Computer-Aided System," 986.

75. The inclusion of amplitude information was based on the assumption that stronger signals had likely followed a more direct path, while oblique signals were attenuated during their longer transits. Since vertical information was of interest, the amplitude information was deemed critical.

76. Lockwood, "A Computer-Aided System," 986.

77. Ibid.

78. Ibid., 988 (emphasis added).

79. Ibid.

80. Galison, *Image and Logic*, 370.

81. Lockwood, "A Computer-Aided System," 988.

82. Jackson, *Alouette-ISIS Program Summary*, 15.

83. McDougall, *The Heavens and the Earth*.

84. Ibid.

Chapter 6

1. Nathaniel C. Gerson, "SIGINT in Space," *Physics in Canada* 54, no. 6 (1998): 353–357.

2. Kurt F. Jensen, *Cautious Beginnings: Canadian Foreign Intelligence, 1939–51* (University of Washington Press, 2009), 7.

3. Martin Rudner, Canada's Communications Security Establishment: From Cold War to Globalization, Occasional Paper 22, Norman Patterson School of International Affairs, Carleton University, 2000, 9–10.

4. Gerson, "SIGINT in Space," 359.

5. The NSA used the words "TOP SECRET" followed by a single code word to classify the sensitivity of its documents. The code word was itself secret and was changed once compromised. Gerson's work was originally classified DAUNT, the highest classification of signals intelligence, for the period from July 1, 1959 through December 1960). UMBRA was adopted December 1, 1968 and was used through October 1999. See *United States District Court for the District of Columbia, Citizens Against Unidentified Flying Objects Secrecy v. National Security Agency*, October 9, 1980, 2.

6. Nathaniel C. Gerson, "Ionospheric Propagation," *NSA Technical Journal* III, no. 4 (1958).

7. On these relationships, see John Cloud, "Imaging the World in a Barrel: CORONA and the Clandestine Convergence of the Earth Sciences," *Social Studies of Science* 31, no. 2 (2001): 231–251.

8. See, for example, Janet Abbate, *Inventing the Internet* (MIT Press, 2000).

9. Andrew B. Godefroy, *Defence and Discovery: Canada's Military Space Program, 1945–74* (University of British Columbia Press, 2012), 26.

10. The other was the German physicist Georg Joos. See Nathaniel C. Gerson, "Collaboration in Geophysics—Canada and the United States 1948–1955," *Physics in Canada* 40, no. 1 (2002): 3–8.

11. Ibid.

12. Nathaniel C. Gerson, "Nathaniel ("Nate") C. Gerson, 1915–2002," *Physics in Canada* 58, no. 2 (2002): 42.

13. Gerson, "Collaboration in Geophysics," 7.

14. Ibid. The radar work was connected to an applied research program in electronics at the DRB for use in the Canadian armed services. See John N. Vardalas, *The Computer Revolution in Canada: Building National Technological Competence* (MIT Press, 2001), 52.

15. Gerson, "Collaboration in Geophysics," 6.

16. In the March 1963 issue of *IEEE Transactions on Radio Frequency Interference*, for example, he is listed as a consultant for ARPA.

17. Nathaniel C. Gerson, "50 Years with Uncle Sam," *Cryptolog* 16, no. 3 (1989): 3–6.

18. Nathaniel C. Gerson, "Six Point Program for Improved Intercept (Part I)," *NSA Technical Journal* VII, no. 3 (1962).

19. Nathaniel C. Gerson, "Antipodal Propagation," *NSA Technical Journal* IV, no. 1 (1959): 55–66.

20. Gerson, "SIGINT in Space," 356.

21. Nathaniel C. Gerson, "Intercept of USSR Missile Transmissions," *NSA Technical Journal* IV, no. 3 (1959): 43–51.

22. National Bureau of Standards, "Forward Scatter of Radio Waves," *Navigation* 5, no. 2 (1956): 107–113.

23. Albert M. Skellett, "The Effect of Meteors on Radio Transmission through the Kennelly-Heaviside Layer," *Physical Review* 37 (1931): 1668; Skellett, "The Ionizing Effect of Meteors," *Proceedings of the Institute of Radio Engineers* 23 (1935): 132–149.

24. Edward Appleton and R. Naismith, "The Radio Detection of Meteor Trails and Allied Phenomena," *Proceedings of the Physical Society* 59 (1947): 461–473; James S. Hey and G. S. Stewart, "Radar Observations of Meteors," *Proceedings of the Physical Society* 59 (1947): 858; Bernard Lovell, *Meteor Astronomy* (Clarendon, 1954), 23–24.

25. Jon Agar, "Making a Meal of the Big Dish: The Construction of the Jodrell Bank Mark 1 Radio Telescope as a Stable Edifice, 1946–57," *British Journal for the History of Science* 27, no. 1 (1994): 3–21; Antony Hewish, "James Stanley Hey, M.B.E. 3 May 1909–27 February 2000," *Biographical Memoirs of Fellows of the Royal Society* 48 (2002): 167–178.

26. Quoted in Andrew J. Butrica, *To See the Unseen: A History of Planetary Radio Astronomy* (NASA History Office), 15. Also see A. C. Lovell, J. A. Clegg, and C. D. Ellyett, "Radio Echoes from the Aurora Borealis," *Nature* 160 (1947): 372; A. Aspinall and G. S. Hawkins, "Radio Echo Reflections from the Aurora Borealis," *Journal of the British Astronomical Association* 60 (1950): 130–135.

27. D. W. R. McKinley, "Meteor Velocities Determined by Radio Observations," *Astrophysics Journal* 113 (1951): 225–267. On the Canadian connection to the British research, see Richard Jarrell, "The Formative Years of Canadian Radio Astronomy," *Journal of the Royal Astronomical Society of Canada* 91 (1997): 20–27.

28. Peter A. Forsyth, William Petrie, and Balfour W Currie, "Auroral Radiation in the 3,000-Megacycle Region," *Nature* 164, no. 4167 (1949): 453.

29. In fact, early meteor researchers at Stanford in 1946 detected meteor ion trails by listening for interference in HF radios tuned to specific frequencies. See Robert A. Helliwell, *Whistlers and Related Ionospheric Phenomena* (Stanford University Press, 1965), 11–23.

30. Thomas L. Eckersley, "Studies in Radio Transmission," *Journal of the Institution of Electrical Engineers* 71, no. 429 (1932): 405–459.

31. Balfour W. Currie, Peter A. Forsyth, and F. E. Vawter, "Radio Reflections from Aurora," *Journal of Geophysical Research* 58 (1953): 179–200.

32. P. A. Forsyth et al., "The Principles of JANET: A Meteor-Burst Communication System," *Proceedings of the IRE* 45, no. 12 (1957): 1642–1657.

33. Forsyth et al., "The Principles of JANET," 1644.

34. See Von Russel Eshleman, *The Mechanism of Radio Reflections from Meteoric Ionization* (Report No. 39, Stanford University Electronics Research Laboratory, 1952); Von Russell Eshleman and Laurence A. Manning, "Radio Communication by Scattering From Meteoric Ionization," *Proceedings of the IRE* 42, no. 3 (1954): 530–536.

35. Forsyth et al., "The Principles of JANET."

36. Ibid.

37. Jens Ostergaard, *Meteor Scatter Communication Between Thule and Station Nord, Greenland* (Hanscom Air Force Base, 1990), 2.

38. Letter from P. A. Forsyth to Frank T. Davies, September 27, 1973, 6, Frank Davies Papers.

39. Paul Baran, *On Distributed Communications*, volume V: *History, Alternative Approaches, and Comparisons* (RAND Corporation, 1964), 2.

40. Baran's system emphasized heuristic routing doctrines to ensure this. See Baran, *On Distributed Communications*, v. Also see Abbate, *Inventing the Internet*, chapter 1.

41. G. W. L. Davis et al., "The Canadian JANET System," *Proceedings of the IRE* 45, no. 12 (1957): 1666–1678; Forsyth et al., "The Principles of JANET," 1656.

42. Ostergaard, *Meteor Scatter Communication Between Thule and Station Nord, Greenland*, 2.

43. Forsyth et al., "The Principles of JANET," 1644.

44. Nathaniel C. Gerson, "High Latitude HF Communications," *IEEE Transactions on Communications Systems* 12, no. 1 (1964): 107–109.

45. Nathaniel C. Gerson, "Radio-Wave and Alternative Communications in the Arctic," *Arctic* 15, no. 3 (1962): 224–228; Gerson, "Radio Wave and Alternate Communications in the Arctic," *Electrical Engineering* 81, no. 5 (1962): 364–366.

46. Gerson, "High Latitude HF Communications," 107.

47. S. C. Corontini, G. E. Hill, and R. Penndorf, "The Problem of Arctic Communications Following Solar Disturbances," in *The Effect of Disturbances of Solar Origin on Communications: Papers Presented at the Symposium of the Ionospheric Research Committee, AGARD Avionics Panel, Naples, Italy*, ed. George J. Gassman, volume AGARDograph 59 (Macmillan, 1963).

48. Baran, *On Distributed Communications*, 5.

49. Gerson, "High Latitude HF Communications," 108–109.

50. Gerson, "Radio Wave and Alternate Communications in the Arctic," 364.

51. This was the 440L system. The transmitters were located in the Philippines, Japan, and Okinawa; receivers were located in Italy, Germany, and England, with a correlation center in Aviano, Italy connected to NORAD Combat Operations. On the attempt to expand these facilities in the late 1960s under Project Clear Sky, see Dean Rusk, State Department Airgram to US Embassy Rome CA-6065, "Project Clear Sky," February 26, 1968; US National Archives, RG 59, US Department of State, Central Foreign Policy files, 1967–1969, file DEF 18–8 US.

52. W. L. Hatton, "Oblique-Sounding and HF Radio Communication," *IRE Transactions on Communications Systems* 9, no. 3 (1961): 275–279.

53. W. L. Hatton, "Survey of the Effects of Ionospheric Propagation on HF Communication Systems," in *Ionospheric Radio Communications*, ed. Kristen Folkestad (Springer, 1968), 210.

54. A number of DRTE researchers worked with this technique. See J. H. Meek, "Oblique Reflection of Radio Waves by Way of a Triangular Path," *Nature*, volume 169 (February 1952), 327; J. Cox and K. Davies, "Oblique Incidence Pulse Transmission," *Wireless Engineer* 32 (February 1955), 35–41; J. H. Chapman, Kenneth Davies, and C. A. Littlewood, "Radio Observations of the Ionosphere at Oblique Incidence," *Canadian Journal of Physics* 33, no. 12 (1955), 713–722; W. L. Hatton and E. E. Stevens, "Oblique Sounding: A Communication Aid," September 29, 1965, 1, Frank Davies Papers; R. G. Gould and W. R. Vincent, System Concepts for a Common-User Radio Transmission Sounding System (CURTS) (Defense Technical Information Center, Stanford Research Institute, 1962).

55. W. L. Hatton, "History of CHEC," 1973, 1, Frank Davies Papers.

56. Ibid. On the technical relationship between oblique sounding and HF channel evaluation, see Hatton, "Survey of the Effects of Ionospheric Propagation on HF Communication Systems," 210.

57. J. C. W. Scott, "Abstract of Brief to Management Committee: Evolution of DRTE Research Programme," January 29, 1958, 1, LAC, RG 24, volume 9, accession 90–91/217, file DRBS 100–22/0, part 2.

58. Hatton, "Oblique-Sounding and HF Radio Communication," 274.

59. W. L. Hatton, "Impact of Information Theory on Future Communication Systems," paper presented at 12th DRB Symposium, Ottawa, November 21, 1960.

60. E. E. Stevens, "World Wide Oblique Incidence Ionospheric Sounding: An Appreciation of the CURTS Program as it applies to Canadian Forces," June 1, 1966, 1, LAC, RG 24, volume 24026, file 2800–30, part 1.

61. Gould and Vincent, System Concepts for a Common-User Radio Transmission Sounding System.

62. Jacob Darwin Hamblin, "The Navy's "Sophisticated" Pursuit of Science," *Isis* 93, no. 1 (2002): 1–27.

63. CANADAIR CP-107 ARGUS 2. http://casmuseum.techno-science.ca/en/collection -research/artifact-canadair-argus-2.php.

64. D. F. Page and W. D. Hindson, "The CHEC System—Towards Automatic Selection of Optimum Communications Channels," *Canadian Aeronautics and Space Journal* 13, no. 7 (1967): 303–306.

65. This was particularly urgent in a region known as the Canlant area—a prime location for Soviet submarine attacks extending roughly from the western tip of Nova Scotia south to the 40th parallel, east along the line of latitude and then north to the southern tip of Greenland.

66. Chief of the Air Staff, Top Secret, "Memorandum to the Chiefs of Staff Committee: Acceleration of RCAF Communications Improvement Program," August 15, 1961, 2, LAC, RG 24, volume 18053, file 952–116–1, part 1.

67. E. W. Pierce, Secret Memorandum, "Radio Communications," May 10, 1965, 2, LAC, RG 24, volume 24026, file 2800–30, part 1.

68. George W. Jull, "H.F. Propagation in the Arctic," in *Arctic Communications: Proceedings of the Eighth Meeting of the AGARD Ionospheric Research Committee, Athens, Greece* (Macmillan, 1964), 157.

69. Nathaniel C. Gerson, "Polar Communications," in *Arctic Communications: Proceedings of the Eighth Meeting of the AGARD Ionospheric Research Committee, Athens, Greece* (Macmillan, 1964).

70. Lew Hatton, "History of CHEC," 1973, 1, Frank Davies Papers.

71. Page and Hindson, "The CHEC System—Towards Automatic Selection of Optimum Communications Channels," 303.

72. Ibid.

73. Ibid., 303, 306; E. E. Stevens, "The CHEC Sounding System," *Ionospheric Radio Communications* (Plenum, 1968), 364.

74. Confidential Letter from Frank Davies to Chief of Defence Staff, April 14, 1966, LAC, RG 24, volume 24026, file DRBS 2800–30, part 1.

75. E. E. Stevens, "Some Personal Recollections of CHEC," October 8, 1973, 2–3, Frank Davies Papers.

76. Stevens, "The CHEC Sounding System" 360.

77. The CHEC system would be adopted, in modified form, by the British Royal Air Force.

78. Secret Memorandum from R. E. Mooney to DG Comm, May 21, 1965, LAC, RG 24, volume 24026, file DRBS 2800–30, part 1.

79. Gerson, "Radio-Wave and Alternative Communications in the Arctic," 228.

Chapter 7

1. On these attempts in governability, see Peter Keith Kulchyski and Frank J. Tester, *Kiumajut (Talking Back): Game Management and Inuit Rights, 1900–70* (University of British Columbia Press, 2007); Caroline Desbiens, *Power From the North: Territory, Identity, and the Culture of Hydroelectricity in Quebec* (University of British Columbia Press, 2014); Hans M. Carlson, *Home Is the Hunter: The James Bay Cree and Their Land* (University of British Columbia Press, 2008); Liza Piper, *The Industrial Transformation*

of Subarctic Canada (University of British Columbia Press, 2010); Matthew Farish and Whitney Lackenbauer, "High Modernism in the Arctic: Planning Frobisher Bay and Inuvik," *Journal of Historical Geography* 35 (2009): 517–544; Lackenbauer and Farish, "The Cold War on Canadian Soil"; John Sandlos and Arn Keeling, "Claiming the New North: Development and Colonialism at the Pine Point Mine, Northwest Territories, Canada," *Environment and History* 18, no. 1 (2012): 5–34; Arn Keeling and John Sandlos, "Environmental Justice Goes Underground? Historical Notes from Canada's Northern Mining Frontier," *Environmental Justice* 2, no. 3 (2009): 117–125.

2. This was different than the use of radio in mapping or the mapping of radio navigation networks. On radio for mapping, see J. E. R. Ross, "Shoran Triangulation in Canada," *Bulletin Géodésique* 24, no. 1 (1952): 207–242; J. E. R. Ross, "Canadian Shoran Effort, 1949–1953," *Empire Survey Review* 12, no. 92 (1954): 242–254. On radio navigation, see Rankin, "The Geography of Radionavigation and the Politics of Intangible Artifacts."

3. Marionne Cronin, "Northern Visions: Aerial Surveying and the Canadian Mining Industry, 1919–1928," *Technology and Culture* 48, no. 2 (2007): 303–330; Stephen Bocking, "A Disciplined Geography: Aviation, Science, and the Cold War in Northern Canada, 1945–1960," *Technology and Culture* 50, no. 2 (2009): 265–290.

4. William Rankin, "The Geography of Radionavigation and the Politics of Intangible Artifacts," *Technology and Culture* 50, no. 3 (2014): 664.

5. For an overview of histories of medical geographies, see Conevery B. Valencius, "Histories of Medical Geography," in *Medical Geography in Historical Perspective*, ed. Nicolaas A. Rupke (Wellcome Institute Trust for the History of Medicine, 2001); Gregg Mitman and Ronald L. Numbers, "From Miasma to Asthma: The Changing Fortunes of Medical Geography in America," *History and Philosophy of the Life Sciences* 25, no. 3 (2003): 391–412; Nicolas Rupke, ed., *Medical Geography in Historical Perspective* (Wellcome Trust Centre for the History of Medicine, 2000); Frank A. Barrett, "Alfred Haviland's Map Analysis of the Geographical Distribution of Diseases in England and Wales," *Social Science and Medicine* 46, no. 6 (1998): 767–781; Saul Jarcho, "Yellow Fever, Cholera, and the Beginnings of Medical Cartography," *Journal of the History of Medicine and Allied Sciences* 25, no. 2 (1970): 131–142; Kari S. McLeod, "Our Sense of Snow: The Myth of John Snow in Medical Geography," *Social Science and Medicine* 50 (2000): 923–935; Marcos Cueto, "Nationalism, Carrión's Disease and Medical Geography in the Peruvian Andes," *History and Philosophy of the Life Sciences* 25, no. 3 (2003): 319–335. John Snow's cholera map of 1854 is one of the earliest examples of health geography in this tradition. See Peter Vinten-Johansen, *Cholera, Chloroform, and the Science of Medicine: A Life of John Snow* (Oxford University Press, 2003), chapter 8.

6. McLeod, "Our Sense of Snow," 163.

7. See Valencius, "Histories of Medical Geography"; Mitman and Numbers, "From Miasma to Asma."

8. See Ramsay Cook, "Loyalism, Technology and Canada's Fate," *Journal of Canadian Studies/Revue d'Études Canadiennes* 5, no. 3 (1970): 50–61.

9. "A Brief History of the NWT & Y Radio System, R C Sigs," LAC, RG 12, volume 16, file 6800–76–1.

10. On the historical demographics of the Yukon, see Kenneth Coates, *Canada's Colonies: A History of the Yukon and Northwest Territories* (Lorimer, 1985), 190–212.

11. For a discussion of the "New North" initiative that framed this integration, see Andrew Stuhl, "The Politics of the 'New North': Putting History and Geography at Stake in Arctic Futures." *Polar Journal* 3, no. 1 (2013): 94–119; Frank J. Tester and Peter Keith Kulchyski, *Tammarniit (Mistakes): Inuit Relocation in the Eastern Arctic, 1939–63* (University of British Columbia Press, 1994).

12. Letter from Jean Lesage to J. J. McCann, September 21, 1954, LAC, RG 85, volume 1210, file 340–1, part 4. In 1949 the Soviet Union had begun to jam Voice of America and BBC broadcasts; the US and Britain had increased the power of their transmissions in an attempt to break through the jamming to reach the Soviet Union and Eastern Europe. By 1954, the powerful transmissions covered the Canadian North, but officials had only a vague idea of how the receptions were distributed across the region and therefore how to combat them.

13. For instance, RPL reports in 1950 included maps titled "Canada—extent of territory and communications," "Canada—population distribution and radio needs," and "Canada—auroral and ionospheric research."

14. The Bal Tabarin was an early-twentieth-century Parisian cabaret in the ninth arrondissement, popular with German officers during World War II and eventually bought by the owners of the Moulin Rouge.

15. Andrew Cowan and William Roxburgh, "Consideration of broadcasting coverage for the Yukon and Northwest Territories," December 12, 1955, 17, LAC, RG 41, volume 127, file 5–2.

16. Ibid.

17. "Report on a Special Meeting called at the Request of Mr. H. Low," LAC, RG 12, volume 1631, file 6800–30, volume 4, 2.

18. On the move of Canadian geography to the laboratory, see Bocking, "A Disciplined Geography." On the Canadian grid method, see Cronin, "Northern Visions."

19. According to the 1951 Census, the population of the Yukon was 9,096 and that of the Northwest Territories was 11,400.

20. The Royal Commission had been convened to investigate an apparent conflict in the CBC's mandate, namely its role as both national broadcaster and national regulator in the 1950s.

21. Gordon Robertson, "Radio in the Northwest Territories," April 1956, 5, 10, LAC, RG 41, volume 127, file 5–2. On Robertson's larger vision Northern development, see Stuhl, "The Politics of the 'New North'"; Tester and Kulchyski, *Tammarniit (Mistakes)*; Matthew Farish and Whitney Lackenbauer, "High Modernism in the Arctic: Planning Frobisher Bay and Inuvik," *Journal of Historical Geography* 35 (2009): 517–544.

22. Robertson, "Radio in the Northwest Territories," 11.

23. Matthew Farish, "Frontier Engineering: From the Globe to the Body in the Cold War Arctic," *The Canadian Geographer/Le Géographe canadien* 50, no. 2 (2006): 177–196. Also see Farish and Lackenbauer, "High Modernism in the Arctic."

24. Robertson, "Report on Radio Broadcasting," 11.

25. F. H. Collins, "Radio in the Yukon Territory: A Brief Presented to the Royal Commission on Broadcasting," April 1956, 4, LAC, RG 41, volume 127, file 5.

26. Canadian Broadcasting Corporation, "Broadcasting Service in Northern Canada," Royal Commission on Broadcasting, 1956, LAC, RG 85, volume 1243, file 340–1, part 7–6.

27. Although not many Inuit and Native families were expected to own a short-wave receiver, Robertson explained that such receivers could be made publicly available and that activities could be organized around them and supplemented by distance education. Similar systems, Robertson noted, were used to communicate with Native communities in sparsely populated parts of Northern Australia and of New Zealand.

28. Andrew Cowan, "Report on Radio Broadcasting in the Yukon and Mackenzie District of the Northwest Territories," December 12, 1955, 4, LAC, RG 41, volume 127, file 5–2.

29. Confidential Letter from Charles Foulkes to R. G. Robertson, October 29, 1956, 2, LAC, RG 12, volume 1621, file 6800–76–1.

30. Andrew Cowan, "Canadian Broadcasting Corporation: Northern Service," 6, LAC, RG 41, volume 127, file 5–2.

31. J. Alphonse Ouimet, "Canada's New Voice: Engineering Features of C.B.C.'s International Shortwave Station," *Engineering Journal* 28 (1945): 440–447.

32. CBC Radio Press Release, "New CBC Shortwave Service Designed for Northern Canada," no. 275, Sept. 2, 1960, 1, LAC, RG 41, volume 128, file 5–4.

33. Quoted in CBC Annual Report, 1958–1959, 1, LAC, RG 85, volume 1230, file 1003–7, part 6.

34. Canadian Broadcasting Corporation, "Northern Service," April 1963, 1, LAC, RG 85, volume 145, file 1003–17, part 9 (emphasis added).

35. CBC, "CBC Northern Service," June 22, 1965, 1, LAC, RG 41, volume 128, file 5–4.

36. "Outline Statement by CBC Northern Service on Co-operation with Government Departments and other Agencies in the Production of Expanded Eskimo and Indian Broadcasts," June 28, 1963, LAC, RG 85, volume 145, file 1003–17, part 9.

37. Canadian Broadcasting Corporation, "Northern Service," April 1963, 1, LAC, RG 85, volume 145, file 1003–17, part 9.

38. CBC, "CBC Northern Service," June 22, 1965, 1, LAC, RG 41, volume 128, file 5–4.

39. Andrew Cowan, "The Medium and the Message," Third Northern Resources Conference, Whitehorse, Y.T., April 10, 1969, 4, LAC, RG 41 volume 127, file 5 (Part 7).

40. Disraeli examined this theme in his Young England novels, particularly *Sybil*.

41. Cowan, "The Medium and the Message," 4. For another expression of this view, see Anne MacLennan, "Cultural Imperialism of the North? The Expansion of the CBC Northern Service and Community Radio," *Radio Journal* 9, no. 1 (2011): 63–81.

42. Department of Plans Memorandum No. 3/62, "Further DRB Participation in Satellite Design and Development," August 24, 1962, 1, LAC, RG 24, accession 83–84/167, volume 7312, file DRBS 0200–501, part 1.

43. John H. Chapman, "Address to General Assembly of URSI in Munich," September 7, 1966.

44. John H. Chapman et al., *Upper Atmosphere and Space Programs in Canada* (Ottawa: Science Secretariat, 1967), 109.

45. In the Northern Hemisphere, the 78th parallel is the world's southernmost integral parallel that does not pass through a continental mainland.

46. Draft Submission to United Nations Committee on Peaceful Uses of Outer Space: Canada-Sweden Project Broadcasting from Satellites, Part III: Technical Considerations, Issue 1, Nov 1966, 1.

47. *Preliminary Report of the Royal Commission on Bilingualism and Biculturalism* (Ottawa, Queen's Printer, 1965), foreword, 5.

48. Symphonie would enable Quebec to receive radio and television programs beamed directly from France, itself eager to avoid US domination of satellite broadcasting and to encourage Quebec nationalism.

49. Letter from John H. Chapman to D. H. W. Kirkwood, September 20, 1967, LAC, RG 97, volume 69, file 4.

50. See Edward Jones-Imhotep, "Paris-Montreal-Babylon: The Modernist Genealogies of Gerald Bull," in *Science, Technology and the Modern in Canada*, ed. Edward Jones-Imhotep and Christina Adcock (University of British Columbia Press, forthcoming); Kevin Toolis, "The Man Behind Iraq's Supergun," *New York Times*, August 26, 1990.

51. On the modernization of government agencies, see Laurence B. Mussio, *Telecom Nation: Telecommunications, Computers, and Governments in Canada* (McGill-Queen's University Press, 2001).

52. For the detailed background to these developments, see Vera Pavri, "What You Say Is What You Get: Policy Discourse and the Regulation of Canada's First Domestic Communications Satellite System," *Technology and Culture* 50, no. 3 (2009): 569–585.

53. Draft of memo from Minister of National Defence to Minister of Communications regarding transfer of DRTE to DOC—November 4, 1968, LAC, RG 97 Series/ Accessions: D.V. Records of the Assistant Deputy Minister's Office (Space Research), 1965–1980, volume 69, file 19.

54. Quoted in Mussio, *Telecom Nation*, 32.

55. Catherine Jutras, "Promises All But Forgotten in Name of Corporate Profit," *Ottawa Citizen*, November 9, 1974.

56. These specific claims were mounted in 1970 by a research team at York University studying the potential impact of Anik. See Jeff Caruthers, "North Skeptical of Satellite TV Plan," *Globe and Mail* (Toronto), March 1, 1972.

57. Louis-Edmond Hamelin, "Essai de Régionalisation du Nord Canadien," *North* (1964): 16–19.

58. In 1964, Hamelin listed his initial criteria: "Nous avons retenu dix critères; 1 - latitude; 2 - gélisol; 3 - nombre de jours dont la temperature est au-dessus de 42°F (S°SC); 4 - indice thermique négatif a partir de 65°F (18°3C); 5 - durée de l'englacement; 6 - couvert vegetal; 7 - communications; 8 - population (indigène et blanche); 9 - exploitation des ressources; 10 - coût des fournitures." See ibid.

59. Although Hamelin's final system was not presented until the 1970s, there is evidence that the underlying principles of nordicity and its various criteria were worked out in the late 1950s and the early 1960s.

60. Louis-Edmond Hamelin and William Barr, *Canadian Nordicity: It's Your North Too* (Harvest House, 1978), xi.

61. *Ecumene* was originally a Greco-Roman term used to denote the inhabited universe. It was later used in cartography to refer to ancient and medieval world maps and in Christianity to denote a unified Christian Church. Lewis Mumford used it in *Technics and Civilization* (1934) to describe world civilizations.

62. Quoted in Coates, *Canada's Colonies*, 215.

63. Hamelin and Barr, *Canadian Nordicity*, xii. Hamelin would define the infelicitous term "Canadianity" as follows: "the identifying traits of Canada. Suggests a state of mind, a mentality, a maturation. To some degree the youth of the country accounts for its limited Canadianity. Different from canadianization, which applies to the processes involved."

64. He expressed the new criteria as follows: "Taking advantage of polar research carried out during the last twenty-five years and of critical comments made by colleagues, I have decided on a family of ten significant, converging factors, which are relevant to the major northern situations: latitude, summer heat above 5.6°C (42°F), annual cold below 0°C (32°F), types of ice (in the ground, on land, or on water), total precipitation, development of the vegetation cover, accessibility (air or surface), number of inhabitants or regional population density, and finally, degree of economic activity." See ibid.

65. Rankin, "The Geography of Radionavigation," 664.

66. On the more general misalignment between the geography of intangible artifacts and conventional political geographies, see ibid.

Conclusion

1. Marshall Berman, *All That Is Solid Melts into Air: The Experience of Modernity* (Viking, 1988), 13.

2. Gregory K. Clancey, *Earthquake Nation: The Cultural Politics of Japanese Seismicity, 1868–1930* (University of California Press, 2006); Matthew Farish, *The Contours of America's Cold War* (University of Minnesota Press, 2010); Gabrielle Hecht, "Enacting Cultural Identity: Risk and Ritual in the French Nuclear Workplace," *Journal of Contemporary History* 32, no. 4 (1997): 483–507.

3. Farish, *Contours of America's Cold War*; P. Whitney Lackenbauer and Matthew Farish, "The Cold War on Canadian Soil: Militarizing a Northern Environment," *Environmental History* 12, no. 4 (2007): 920–950.

4. See Holger Nehring, "What Was the Cold War?" *English Historical Review* 127, no. 527 (2012): 920–949.

5. On "states of emergency," see Minami Orihara and Gregory Clancey, "The Nature of Emergency: The Great Kanto Earthquake and the Crisis of Reason in Late Imperial Japan," *Science in Context* 25, no. 1 (2012): 103–126.

6. See the collection of essays on "Colonial Science" in *Isis* 96, no. 1 (2005): 52–87; John Law, "Technology and Heterogeneous Engineering: The Case of Portuguese Expansion," in *The Social Construction of Technological Systems: New Directions in the Sociology and History of Technology*, ed. Wiebe E. Bijker, Thomas P. Hughes, and Trevor Pinch (MIT Press, 1987); Simon Schaffer, "Easily Cracked: Scientific Instruments in States of Disrepair," *Isis* 102, no. 4 (2011): 706–717.

7. These complex technologies and their dangers provided archetypes for the "risk society." See Ulrich Beck, *Risk Society: Toward a New Modernity* (SAGE, 1992); Anique Hommels, Jessica Mesman, and Wiebe E. Bijker, eds., *Vulnerability in Technological Cultures: New Directions in Research and Governance* (MIT Press, 2014); Deborah R. Coen, *The Earthquake Observers: Disaster Science from Lisbon to Richter* (University of Chicago Press, 2012), 268; Edward Jones-Imhotep, "Disciplining Technology: Electronic Reliability, Cold-War Military Culture and the Topside Ionogram," *History and Technology* 17, no. 2 (2000): 125–175; Charles Perrow, *Normal Accidents: Living with High-Risk Technologies* (Princeton University Press, 1999); David Nye, *When the Lights Went Out: A History of Blackouts in America* (MIT Press, 2010).

8. On the vanishing of disasters, see Coen, *The Earthquake Observers*, 276.

9. See Bruno Latour, "Give Me a Laboratory and I Will Raise the World," in *Science Observed: Perspectives on the Social Study of Science*, ed. Karin Knorr-Cetina and Michael Mulkay (SAGE, 1983); Timothy Mitchell, *The Rule of Experts: Egypt, Techno-Politics, Modernity* (University of California Press, 2002), chapter 3. On regimes of perceptibility, see Michelle Murphy, *Sick Building Syndrome and the Problem of Uncertainty: Environmental Politics, Technoscience, and Women Workers* (Duke University Press, 2006), 12.

10. See, for example, Robert Brain, "Standards and Semiotics," in *Inscribing Science: Scientific Texts and the Materiality of Communication*, ed. Timothy Lenoir (Stanford University Press, 1997); Valencius, "Histories of Medical Geography," in *Medical Geography in Historical Perspective*, ed. Nicolaas A. Rupke (Wellcome Institute Trust for the History of Medicine, 2001).

11. On the invisibility of the social, see Graham and Thrift, "Out of Order"; Simon Schaffer, "Babbage's Intelligence: Calculating Engines and the Factory System," *Critical Inquiry* 21, no. 1 (1994): 203–227; Steven Shapin and Simon Schaffer, *Leviathan and the Air-Pump: Hobbes, Boyle, and the Experimental Life* (Princeton University Press, 1985). On invisibilities in the history of technology, see David Edgerton, *The Shock of the Old: Technology and Global History Since 1900* (Oxford University Press, 2011). On breakage as revealing the invisible, see Martin Heidegger, *The Question Concerning Technology, and Other Essay* (Harper Collins, 1977).

12. Ronald Doel, "Constituting the Postwar Earth Sciences: The Military's Influence on the Environmental Sciences in the USA After 1945," *Social Studies of Science* 33, no. 5 (2003): 635–666; Doel, "Earth Sciences and Geophysics," in *Science in the Twentieth Century*, ed. John Krige and Dominique Pestre (Harwood, 1997); Michael Aaron Dennis, "Postscript: Earthly Matters: On the Cold War and the Earth Sciences," *Social Studies of Science* 33, no. 5 (2003): 809–819; Jacob Darwin Hamblin, "The Navy's "Sophisticated" Pursuit of Science," *Isis* 93, no. 1 (2002): 1–27.

13. On overdetermination in the context of aircraft accidents, see Peter Galison, "An Accident of History," in *Atmospheric Flight in the Twentieth Century*, ed. Peter Galison and Alex Roland (Kluwer, 2000).

14. For similar arguments about the history of science and of technological design, see Lissa Roberts, "Science and Global History, 1750–1850: Local Encounters and Global Circulation," *Itinerario* 33, no. 1 (2009): 7–8; Gabrielle Hecht, "Political Designs: Nuclear Reactors and National Policy in Postwar France," *Technology and Culture* 35, no. 4 (1994): 657–685.

15. See James C. Scott, *Seeing Like a State: How Certain Schemes to Improve the Human Condition Have Failed* (Yale University Press, 1998). On high modernism in the Canadian context, see Tina Loo, "High Modernism, Conflict and the Nature of Change in Canada: A Look at *Seeing Like a State*," *Canadian Historical Review* 97 (2016): 34–58; Daniel Macfarlane, *Negotiating a River: Canada, the US, and the Creation of the St. Lawrence Seaway* (University of British Columbia Press, 2014).

16. See Dipesh Chakrabarty, *Provincializing Europe: Postcolonial Thought and Historical Difference* (Princeton University Press, 2000).

Archival Collections Used

Library and Archives Canada, Ottawa

———. Andrew Gillespie Cowan fonds.

———. John Herbert Chapman fonds.

———. Department of Communications fonds.

———. Department of External Affairs fonds.

———. Department of Indian Affairs and Northern Development fonds.

———. Department of National Defence fonds.

———. Privy Council Office fonds.

Document Collection, Canada Science and Technology Museum, Ottawa

Frank Davies Papers, Communication Research Centre, Ottawa

U.S. Department of State Central Foreign Policy Files, U.S. National Archives, College Park, Maryland

Index

Inside Technology

edited by Wiebe E. Bijker, W. Bernard Carlson, and Trevor Pinch

Edward Jones-Imhotep, *The Unreliable Nation: Hostile Nature and Technological Failure in the Cold War*

Pablo J. Boczkowski and C. W. Anderson, editors, *Remaking the News: Essays on the Future of Journalism Scholarship in the Digital Age*

Benoît Godîn, *Models of Innovation: History of an Idea*

Brice Laurent, *Democratic Experiments: Problematizing Nanotechnology and Democracy in Europe and the United States*

Stephen Hilgartner, *Reordering Life: Knowledge and Control in the Genomics Revolution*

Cyrus C. M. Mody, *The Long Arm of Moore's Law: Microelectronics and American Science*

Harry Collins, Robert Evans, and Christopher Higgins, *Bad Call: Technology's Attack on Referees and Umpires and How to Fix It*

Tiago Saraiva, *Fascist Pigs: Technoscientific Organisms and the History of Fascism*

Teun Zuiderent-Jerak, *Situated Intervention: Sociological Experiments in Health Care*

Basile Zimmermann, *Technology and Cultural Difference: Electronic Music Devices, Social Networking Sites, and Computer Encodings in Contemporary China*

Andrew J. Nelson, *The Sound of Innovation: Stanford and the Computer Music Revolution*

Sonja D. Schmid, *Producing Power: The Pre-Chernobyl History of the Soviet Nuclear Industry*

Casey O'Donnell, *Developer's Dilemma: The Secret World of Videogame Creators*

Christina Dunbar-Hester, *Low Power to the People: Pirates, Protest, and Politics in FM Radio Activism*

Eden Medina, Ivan da Costa Marques, and Christina Holmes, editors, *Beyond Imported Magic: Essays on Science, Technology, and Society in Latin America*

Anique Hommels, Jessica Mesman, and Wiebe E. Bijker, editors, *Vulnerability in Technological Cultures: New Directions in Research and Governance*

Amit Prasad, *Imperial Technoscience: Transnational Histories of MRI in the United States, Britain, and India*

Charis Thompson, *Good Science: The Ethical Choreography of Stem Cell Research*

Nelly Oudshoorn and Trevor Pinch, editors, *How Users Matter: The Co-Construction of Users and Technology*

Peter Keating and Alberto Cambrosio, *Biomedical Platforms: Realigning the Normal and the Pathological in Late-Twentieth-Century Medicine*

Paul Rosen, *Framing Production: Technology, Culture, and Change in the British Bicycle Industry*

Maggie Mort, *Building the Trident Network: A Study of the Enrollment of People, Knowledge, and Machines*

Donald MacKenzie, *Mechanizing Proof: Computing, Risk, and Trust*

Geoffrey C. Bowker and Susan Leigh Star, *Sorting Things Out: Classification and Its Consequences*

Charles Bazerman, *The Languages of Edison's Light*

Janet Abbate, *Inventing the Internet*

Herbert Gottweis, *Governing Molecules: The Discursive Politics of Genetic Engineering in Europe and the United States*

Kathryn Henderson, *On Line and On Paper: Visual Representation, Visual Culture, and Computer Graphics in Design Engineering*

Susanne K. Schmidt and Raymund Werle, *Coordinating Technology: Studies in the International Standardization of Telecommunications*

Marc Berg, *Rationalizing Medical Work: Decision-Support Techniques and Medical Practices*

Eda Kranakis, *Constructing a Bridge: An Exploration of Engineering Culture, Design, and Research in Nineteenth-Century France and America*

Paul N. Edwards, *The Closed World: Computers and the Politics of Discourse in Cold War America*

Donald MacKenzie, *Knowing Machines: Essays on Technical Change*

Wiebe E. Bijker, *Of Bicycles, Bakelites, and Bulbs: Toward a Theory of Sociotechnical Change*

Louis L. Bucciarelli, *Designing Engineers*

Geoffrey C. Bowker, *Science on the Run: Information Management and Industrial Geophysics at Schlumberger, 1920–1940*

Wiebe E. Bijker and John Law, editors, *Shaping Technology / Building Society: Studies in Sociotechnical Change*

Stuart Blume, *Insight and Industry: On the Dynamics of Technological Change in Medicine*

Donald MacKenzie, *Inventing Accuracy: A Historical Sociology of Nuclear Missile Guidance*

Pamela E. Mack, *Viewing the Earth: The Social Construction of the Landsat Satellite System*

H. M. Collins, *Artificial Experts: Social Knowledge and Intelligent Machines*

http://mitpress.mit.edu/books/series/inside-technology